CHAOTIC DYNAMICS IN HAMILTONIAN SYSTEMS

With Applications to Celestial Mechanics

WORLD SCIENTIFIC SERIES ON NONLINEAR SCIENCE

Editor: Leon O. Chua
University of California, Berkeley

Series A. MONOGRAPHS AND TREATISES*

Volume 78: Qualitative and Asymptotic Analysis of Differential Equations with Random Perturbations
A. M. Samoilenko & O. Stanzhytskyi

Volume 79: Robust Chaos and Its Applications
Z. Elhadj & J. C. Sprott

Volume 80: A Nonlinear Dynamics Perspective of Wolfram's New Kind of Science (Volume V)
L. O. Chua

Volume 81: Chaos in Nature
C. Letellier

Volume 82: Development of Memristor Based Circuits
H. H.-C. Iu & A. L. Fitch

Volume 83: Advances in Wave Turbulence
V. Shrira & S. Nazarenko

Volume 84: Topology and Dynamics of Chaos: In Celebration of Robert Gilmore's 70th Birthday
C. Letellier & R. Gilmore

Volume 85: A Nonlinear Dynamics Perspective of Wolfram's New Kind of Science: (Volume VI)
L. O. Chua

Volume 86: Elements of Mathematical Theory of Evolutionary Equations in Banach Spaces
A. M. Samoilenko & Y. V. Teplinsky

Volume 87: Integral Dynamical Models: Singularities, Signals and Control
D. Sidorov

Volume 88: Wave Momentum and Quasi-Particles in Physical Acoustics
G. A. Maugin & M. Rousseau

Volume 89: Modeling Love Dynamics
S. Rinaldi, F. D. Rossa, F. Dercole, A. Gragnani & P. Landi

Volume 90: Deterministic Chaos in One-Dimensional Continuous Systems
J. Awrejcewicz, V. A. Krysko, I. V. Papkova & A. V. Krysko

Volume 91: Control of Imperfect Nonlinear Electromechanical Large Scale Systems: From Dynamics to Hardware Implementation
L. Fortuna, A. Buscarino, M. Frasca & C. Famoso

Volume 92: Modeling, Analysis and Control of Dynamical Systems with Friction and Impacts
P. Olejnik, J. Awrejcewicz & M. Fečkan

Volume 93: Coupled Phase Locked Loops: Stability, Synchronization, Chaos and Communication with Chaos
V. V. Matrosov & V. D. Shalfeev

*To view the complete list of the published volumes in the series, please visit:
http://www.worldscientific.com/series/wssnsa

WORLD SCIENTIFIC SERIES ON NONLINEAR SCIENCE Series A Vol. 25

Series Editor: Leon O. Chua

CHAOTIC DYNAMICS IN HAMILTONIAN SYSTEMS

With Applications to Celestial Mechanics

Harry Dankowicz

Department of Mechanics
Royal Institute of Technology
Stockholm, Sweden

World Scientific
Singapore • New Jersey • London • Hong Kong

Published by

World Scientific Publishing Co. Pte. Ltd.
5 Toh Tuck Link, Singapore 596224
USA office: 27 Warren Street, Suite 401-402, Hackensack, NJ 07601
UK office: 57 Shelton Street, Covent Garden, London WC2H 9HE

Library of Congress Cataloging-in-Publication Data
Dankowicz, H.
Chaotic dynamics in Hamiltonian systems : with applications to celestial mechanics / H. Dankowicz.
p. cm. -- (World Scientific series on nonlinear science. Series A ; vol. 25)
Includes bibliographical references and index.
ISBN-13 978-981-02-3221-4 -- ISBN-10 981-02-3221-7
1. Celestial mechanics. 2. Hamiltonian systems. 3. Chaotic behavior in systems. I. Title. II. Series. World Scientific series on nonlinear science. Series A, Monographs and treatises v. 25.
QB351.D36 1997
531'.11'0151474--dc21 97-33097
CIP

British Library Cataloguing-in-Publication Data
A catalogue record for this book is available from the British Library.

Excerpts from pages 36 and 53 of Bertrand Russell's book *History of Western Philosophy*, © 1946 George Allen and Unwin Ltd, have been included in the preface with kind permission from Routledge Ltd.

Excerpts from the article "Escape of Particles Orbiting Asteroids in the Presence of Radiation Pressure Through Separatrix Splitting," by H. Dankowicz, *Celestial Mechanics and Dynamical Astronomy* **67:1**, 1997, © 1997 Kluwer Academic Publishers have been included in Chapter 7 with kind permission from Kluwer Academic Publishers.

to all the shoemakers in my family

Preface

Science may set limits to knowledge,
but should not set limits to imagination.
Bertrand Russell

The story in this book is a product of the imagination of many minds. People have been occupied with the apparent changing nature of their environment ever since the dawn of humanity. In some instances this occupation took its refuge in theological explanations and metaphysical philosophy. The dynamical aspects of existence were questioned and polemicized from many a perspective, from the skeptics' viewpoint of motion as a deception of the senses, to Heraclitus' claim on change as the fabric of existence. The development of an empirically based scientific method eventually allowed relevant observations to be formalized and the language that arose, mechanics, to be used to further our understanding of motion.

The nature of dynamics, while definitely not understood in a metaphysical sense, has nevertheless yielded to some of science's most ingenious imaginations. The laws of mechanics, as described in the language of mathematics, and in particular, differential equations, have successfully been applied to very large and distant objects, as well as those small and nearby. So successful has this endeavor been that sometimes there has arisen a precarious belief in the "truth" contained in these laws. The laws, in all their simplicity, have been relegated to an ideal world of platonic ideas where they transcend their incomplete incarnations in the present world. Or, in the words of Bertrand Russell in his *History of Western Philosophy*:

> Most sciences, at their inception, have been connected with some form of false belief, which gave them a fictitious value. Astronomy was connected with astrology, chemistry with alchemy. Mathematics was associated with a more refined type of error. Mathematical knowledge appeared to be certain, exact, and applicable to the real world; moreover it was obtained by mere thinking, without the need of observation. Consequently, it was thought to supply an ideal, from which every-day empirical knowledge fell short. It was supposed, on the basis of mathematics, that thought is superior to sense, intuition to observation. If the world of sense does not fit mathematics, so much the worse for the world of sense.

But mechanics is part of the world of sense. Its everpresence in our daily lives makes it an urgent matter to which science in its many guises inevitably finds itself drawn. Dynamical phenomena range from the apparently simple and predictable motion of a vibrating string in a musical instrument to the complex flow of fluid through the heart of an animal. Despite the enormous scope of mechanics, be it the celestial mechanics of stellar bodies or the turbulent flow of air in human lungs, the predictive powers of the mathematical laws of mechanics have completely justified their continued use.

The realization that complex behavior is contained within a seemingly simple mathematical formulation without the need to introduce random effects due to external disturbances has dawned on its investigators primarily in the past hundred years. Human imagination, largely that of Henri Poincaré, has conceived of intuitive tools to describe the origin of such complex dynamics. A fascinating world of possibilities has appeared, a pandorian box of secrets previously unknown. Where science has still only begun to nudge the boundaries of this field, the potential applications are as numerous as they are encouraging.

This book attempts to describe some aspects of the theoretical foundations of the phenomena described above. In the spirit of the previous remarks, the theory attains its remarkable beauty primarily in its interactions with real physical examples. A rich and complicated weave of mathematical ideas truly sparkles only in the light of the stars. This is also the case in this text, where the complex dynamics of small circumasteroidal grains under the perturbation of radiation pressure originating in the sun are studied. Understanding its features is of astronomical interest as well as practical importance, should missions ever to try to approach asteroids at close distances.

Communication is difficult enough between people who share a common language. Imagine then the endless confusion possible in its absence. The language of this book is primarily that of mathematics and, in particular, the theory of dynamical systems. There is probably no replacement for reading it from the beginning if only occasionally skimming where familiar facts are discussed. I have attempted to retain conventional notation and terminology throughout the text, but what is conventional for one may be atypical for another.

This book is not, however, a book of mathematics. It is an attempt to describe a set of complex ideas in enough detail to satisfy the more inquiring minds, for whom ambiguity through a popular treatment falls short. Consequently, many results are not proved from fundamental definitions, but rather further references are made to the existing literature. Nevertheless, it is my hope that a suitable balance has been achieved between the rigor of a mathematical treatment and the phenomenology of intuition.

The intended audience is a first or second year graduate student with a basic background in differential equations and classical mechanics. I believe there is much to interest the applied mathematician as well as the theoretical mechanical engineer and astronomer. Appropriate judgment must be used in determining the best approach to reading this book, depending on one's background and interest.

In the first chapter of this book the leitmotif of each character of the symphony is introduced. The theory is not meant to be complete; instead, only relevant properties are considered and preliminary analysis, as pertains to later chapters, is presented. This chapter covers the geometrical theory of dynamical systems, some basic facts of differential equations, the laws of mechanics and particularly the Hamiltonian formulation, and finally the application to celestial mechanics.

Chapter 2 picks up the tune of Hamiltonian mechanics and describes in detail some techniques associated with its analysis. In particular, the first dissonant chords in the previously regular melody are introduced in connection with the KAM theorem. In particular, the remnants of predictable behavior contained in the KAM result are discussed using methods developed in the previous decade that are alternative to those used in its original proof.

In Chapter 3 we widen our perspective to general dynamical systems and particularly to those satisfying a certain property, that of exponential splittings. The chapter aims at proving the existence of chaotic and highly unpredictable behavior in a simple mathematical system given certain geometric facts. As in the previous chapter, an alternative method of proof to the horseshoe construction usually given in the literature, is provided. Particularly useful are the geometric ingredients in the proof that return again in later chapters. Finally, the numerical tool of Liapunov exponents is introduced as a convenient means of detecting chaos in physical applications.

A perturbation method applicable to proving the presence of chaos in a dynamical system is considered in Chapter 4. The Melnikov method, first conceived in the 1960s and fully developed in the past decade, is presented in detail, with the additional aim of laying the groundwork for chapters to come. The theory is exemplified in a particular application requiring the alternative treatment to standard derivations given in this chapter.

In the fifth chapter the astronomical problem alluded to above is introduced. The physical background is outlined and the methods of the previous chapters are applied to the particularly simple case of planar motion. We argue for the presence of irregular motion in the dynamics of small grains orbiting asteroids and perturbed by radiation pressure. While the consequences are relatively mild in the two-degree-of-freedom case, similar results in Chapter 7 are found to be of possible physical importance.

In fact, Chapter 6 generalizes the methods and ideas of Chapters 4 and 5 to the problem of more degrees-of-freedom, in particular three or higher. The higher-dimensionality of the relevant systems of ordinary differential equations dramatically increases the possibility of irregular behavior, particularly in the form of large excursions in the classical adiabatic invariants. These results are finally applied to the full three-dimensional motion of the circumasteroidal grains in Chapter 7. More specifically, the possibility of grains escaping the vicinity of the asteroid due to radiation pressure is investigated and the relevant time scales are estimated.

This book concludes with a brief outline of some open-ended questions that are at the forefront of research in the areas described in the previous chapters. It is suggested that the reader continually refer to this last section, so as to retain an appropriate sense of incompleteness, rather than, as often is the case in books of mathematics, to be deceived by the complexity of the theory to believe in its completeness.

Finally, a word on the textual layout employed. In particular, I have chosen to intersperse *illustrations* into the running text. These are **not** simply examples of the theory. Instead, they often contain results used in the following text and should thus be read along with the rest of the text.

I find existence fascinating and awesome. It is hard if not impossible to fathom the delicate yet immensely powerful workings in all aspects of this giant process. Equally unfathomable is the apparent structure and susceptibility to the theories of science which nature displays. In all modesty I have written this book to share some remarkable castles in the air with which some of us are preoccupied and which we believe to be useful in describing nature.

There are many books that treat the ideas described here in more or less detail. I have attempted to write a book I would have liked to read, but in the course of its writing my perspective has changed and I am no longer who I was at its conception. It is hard for me to judge the success of my endeavor. Nevertheless, there it is.

Many people have been very helpful in teaching me the ideas contained herein. In particular, Martin Lesser, who awoke my interest in the field of dynamical systems, has continued to encourage me throughout and moreso with respect to the writing of this book. For this, and many other things, I owe him much gratitude. Philip Holmes, my doctoral advisor, guided me and stimulated my quest in dynamical systems. An inspiring man with many-faceted interests in science and an ability to convey this fascination to others, his advice and example deserves my appreciation. Further thanks go to Richard Rand, George Haller, Alessandro Morbidelli, Pierre Lochak, Hanno Essén, Håkan Eliasson, Martin Zimmerman, and many others for valuable discussions and their invaluable knowledge. Many thanks also to my mother-in-law whose technical editing greatly enhanced the readability of this text.

I believe that my writing this book has been fairly inconsequential to most around me and has not perturbed their lives greatly. Everyone, that is, except our parakeet, Jamie, who has had to stay in his cage more often than we would have liked.

Finally, to my wife and mother who are closest to me in my life, for their unconditional support. I love them deeply.

Contents

Chapter 1
INTRODUCTION

I saw Eternity the other night,
Like a great ring of pure and endless light,
All calm, as it was bright;
And round beneath it, Time in hours, days, years,
Driven by the spheres
Like a vast shadow moved; in which the world
And all her train were hurled.

Henry Vaughan

It is a well-trodden path on which we journey. Of the three main subjects of this book–dynamical systems, Hamiltonian dynamics, and celestial mechanics–the last is by far the oldest. The study of the large-scale, internal motions of the solar system goes back to preantiquity. In early civilization (and to this day), the bodies of the solar system were attributed divine origin and their motions were believed to be signs that were interpreted by those versed in the ways of the gods. In ancient cultures, the seemingly regular motions of these celestial objects were a reflection of a harmony of life, with repercussions on one's daily experiences. Irregular behavior was seen mostly as an aberration, impossible to derive structure from and not amenable to predictions.

Religious prejudices and limited means of observation led astronomers to hold on to an earth-centered view of the universe. The heavenly objects were thought to orbit the earth, which in turn remained immobile, freely suspended in the universe. Geometric considerations, coupled with theological beliefs in the perfection of certain ideal structures, led to the conclusion that the celestial bodies, in fact, moved on spheres centered on earth. Where these theories failed to agree with actual observations, complicated geometric constructions were adopted in which several layers of cyclic motions, so-called epicycles, were superposed onto each other. Such a description, first put forward by the Greek astronomer Apollonius in the third century B.C.E., came to be known as the Ptolemaian model, after the second century astronomer Ptolemy. Heavily underwritten by Aristotle, as well as the Catholic church, it survived for almost two millennia and was not finally dethroned until the advent of Newtonian mechanics.

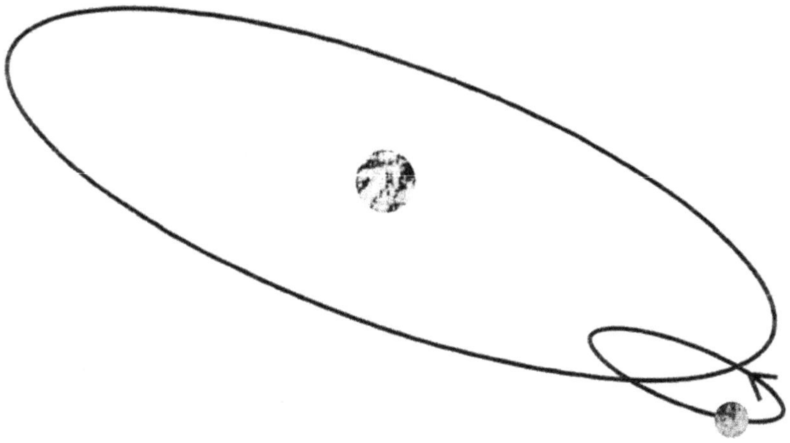

The Ptolemaian model explained the noncircular orbits of the planets around the earth through auxiliary superposed epicycles.

The first stab at the Ptolemaic model was given by Copernicus, whose heliocentric system required the planets to move around the sun. The idea had already appeared in the second century B.C.E. in the teachings of Aristarchus of Samos, but received very little attention and had no impact on contemporary astronomy. The Copernican system, while shifting the focus to motion around the sun, still retained traces of the epicyclic construction. Relying on the belief that circular motion, in its simple form, was fundamental to the motions of celestial bodies, the theory was unable to reconcile predictions and measurements. The observations of Galileo that comets moved on parabolic orbits, and those of Kepler that the planets followed elliptical orbits eventually led to modifications of the Copernican model.

The Keplerian theory is contained in the three Keplerian laws:

1. The planets move on elliptical orbits with the sun at a focus.
2. The motions of the planets are constrained to planes such that the position vector of a planet relative to the sun sweeps out equal areas in equal times.
3. The cube of the semimajor axis goes as the square of the period of the motion.

These laws were the results of deductions made from a great many observations of the motions of the planets by the Danish astronomer and nobleman Tycho Brahe. Toward the end of Brahe's life, Kepler acted as Brahe's assistant and eventually came into possession of Brahe's records.

While only limited observations were available to Kepler, his laws were found to fit the motion of Mars much better than did the older constructions based on epicycles. There was initially no reason to extrapolate from the motion of the planets around the sun to the motion of bodies in general. This process was greatly afforded by Galileo's discovery of four of Jupiter's moons whose orbits were found to agree with the Keplerian laws, albeit with Jupiter taking the place of the sun. This new understanding prepared the ground for the Newtonian theory of gravitation which, together with Newton's laws of motion, was able to explain all of the observed dynamics of the celestial bodies.

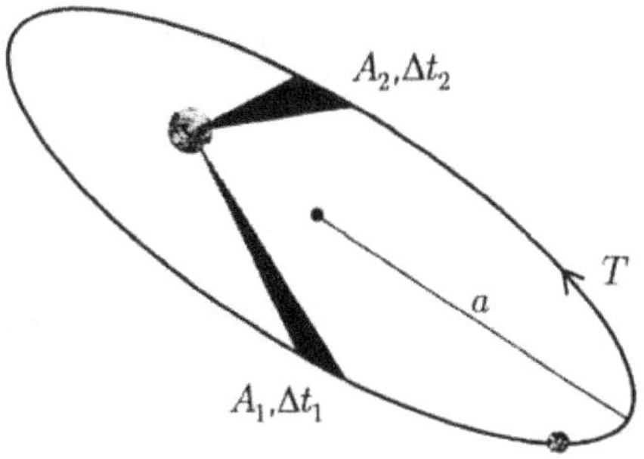

Kepler's laws imply that $A_1 = A_2$ *provided* $\Delta t_1 = \Delta t_2$*, and moreover that* $T^2 \sim a^3$.

The subject of classical mechanics, born in the work of Newton, was brought to maturity in the nineteenth century through the work of many investigators, most prominently Hamilton and Lagrange. The reformulation of the vectorial equations of motion of Newton, through the use of variational principles, proved to have repercussions in all branches of theoretical physics. In celestial mechanics, much was learned about the effects of perturbations on the motion of the planets originating in mutual gravitational interactions. A striking success was the prediction of an additional planet outside the Uranian orbit. While large anomalies between the predicted and observed orbits of Uranus led some investigators to propose modifications of the Newtonian laws, the work of Adams and Leverrier retained Newtonian gravity, while indicating the existence of a perturbing mass beyond Uranus, the planet today known as Neptune.

While the Newtonian equations describing the motion of two mutually attracting bodies were solvable and the conditions for bounded, stable motions were easily found, such results proved elusive in the case of several bodies. Already in the case of three bodies did the established perturbation methods fail to yield conclusive answers. The question of the long-term stability of the solar system received much attention and it was in an effort to find an answer that Poincaré around the turn of the century laid the foundation for the modern theory of dynamical systems.

Combined with the older analytic approach to studying differential and difference equations, the geometric viewpoint expounded by dynamical systems theory has proven suitable for explaining many observed physical phenomena. The intuition afforded by the ability to conjure graphical analogies to analytic concepts lies at the core of many deep results about the behavior of dynamical systems. In its generality, the theory has been applied to a large number of areas, including that in which it originated. In addition to providing partial answers to many of the unanswered questions of celestial mechanics, new and unexpected types of dynamics have been predicted both qualitatively and quantitatively.

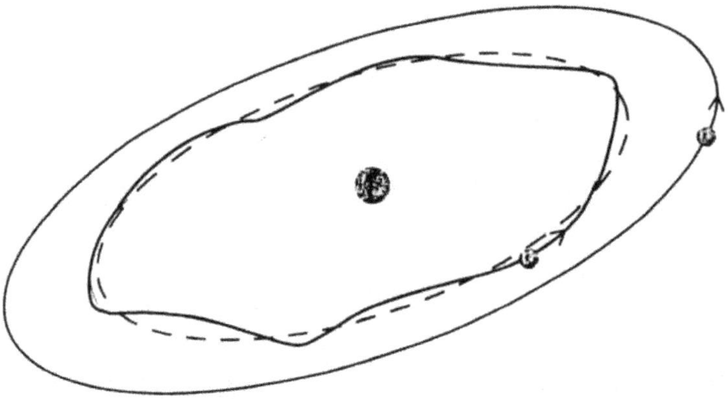

The apparently non-Newtonian motion of Uranus led to the suggestion and later discovery of the existence of Neptune.

This chapter supplies the theoretical background to the chapters that follow. Many of the results are likely well known to some readers, while unfamiliar to others. In an attempt to develop a common vocabulary, as well as provide some fundamental tools for the manipulation of the introduced concepts, the chapter should be enjoyed for its novelty by the novices and its familiarity by the initiated.

Section 1 introduces some basic terminology and perturbation results in the context of a dynamical system and its behavior in state space. As a precursor to Chapter 3, particular emphasis is put on the derivation of inequalities for the dynamical evolution in the presence of a hyperbolic structure. In Section 2, the theory of Hamiltonian systems is presented including some useful results and techniques. We show in Section 3 how classical mechanics can be conveniently formulated in the Hamiltonian context. The Newtonian theory of gravitation and its implications on the fundamental two-body problem of celestial mechanics are described in Section 4, and the stage is set for subsequent developments in the later chapters.

1.1 Dynamical systems

1.1.1 General terminology

We refer to a physical and/or mathematical system consisting of

- a set of **state variables**, whose current values constitute the **state** of the system, and
- a well-defined rule from which the evolution of the state with respect to an independent variable known as **time** can be derived,

as a **dynamical system**. (Often this description refers to the evolutionary rule itself, with the state variables being understood from the context.) It is assumed that no exogenous variables, other than time itself, are necessary to describe this evolution. We refer to the number of variables in the state as the **dimension** of the system. For example, finite-dimensional systems arise in the study of rigid body mechanics, where the positions and velocities of a finite number of points suffice to describe the dynamics of the bodies. On the other hand, a complete description of the mechanics of deformable bodies requires an infinite number of points. These are thus referred to as infinite-dimensional systems. In this text our focus will be on finite-dimensional systems.

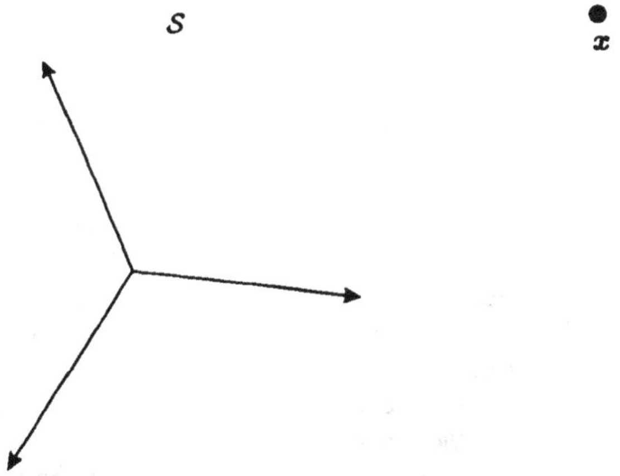

A model state space.

We denote the state of a system using the vector notation, $\boldsymbol{x} = (x_1, ..., x_n)^T$, where n is the dimension of the system. A standard tool for aiding the intuitive interpretation of dynamical systems is the concept of a **state space**, S. This should not be confused with the configuration space of mechanical systems, since a particular configuration corresponds to a multitude of different points in state space, each with different velocity. A particular state simply corresponds to a point in $\mathcal{S}$. Consequently, the dimension of state space equals the dimension of the dynamical system. A set of states, $\{\boldsymbol{x}^t\}$, parameterized by t in some ordered set $\mathcal{I}$ (be it an interval of $\mathbf{R}$ or an increasing sequence in $\mathbf{Z}$), corresponding to the evolving state over an interval of time, is referred to as a segment of an **orbit** of the dynamical system. If we can extend the evolution to all future times, we talk of the **positive orbit** and, similarly, the **negative orbit** for past times. We shall repeatedly use the idea of a state space to illustrate our approach.

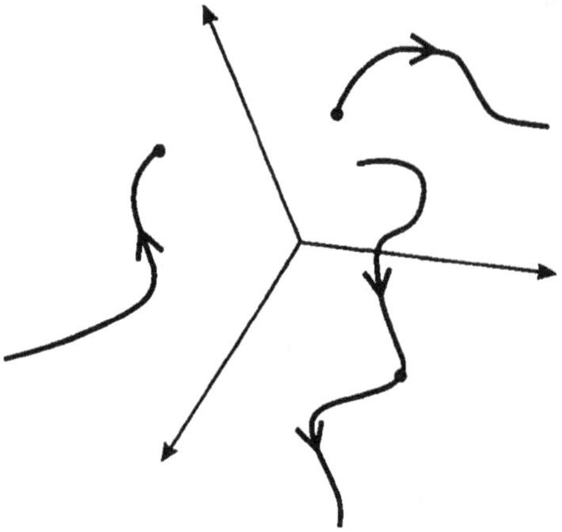

Segments of positive, negative and complete orbits.

Illustration

*A **phase portrait** of a dynamical system is a graphical depiction of the states and their orbits. It is common to omit explicit reference to time other than by putting a little arrow on the orbit to denote the direction of increasing time. We note that, in general, orbits may not exist for all times. Moreover, they may intersect so that the future or the past is not uniquely determined by the present. Even in the presence of a differentiable structure on S, i.e., if S is a manifold, it is not generally true that the orbits will be smooth or even continuous.* •

When the evolution rule is explicitly time-dependent, we speak of a **nonautonomous** system. Similarly, the absence of such dependence indicates an **autonomous** system. By enlarging the state vector to include time as a component, a nonautonomous system can, in principle, be converted to an autonomous one. In particular, given the dynamical system

$$\boldsymbol{x}, \Xi : \mathcal{S} \times \mathcal{I} \to \mathcal{S} \tag{1.1}$$

where Ξ describes the evolution of the state $\boldsymbol{x} \in \mathcal{S}$ for times $t \in \mathcal{I}$, we introduce the **suspended** dynamical system

$$(\boldsymbol{x}, x_{n+1}), \Xi^* : \mathcal{S}^* \to \mathcal{S}^*. \tag{1.2}$$

Here, Ξ^* is identical to Ξ on the first n arguments and essentially takes the current time to the following member in $\mathcal{I}$. This is a natural step if one considers the explicit time-dependence to be the result of influences from external dynamics. Enlarging the set of state variables is then equivalent to replacing the effects of such dynamics by the introduction of an additional variable.

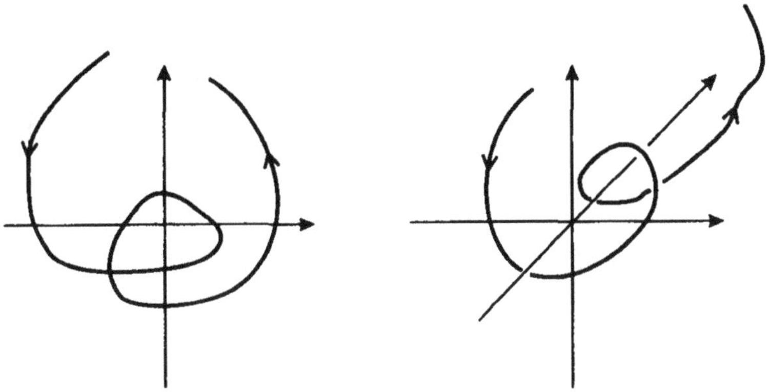

Introducing time as an additional state variable unwraps the previously self-intersecting orbit. (Note that time is assumed to extend into the page.)

Under certain conditions an autonomous system can similarly be converted to a nonautonomous one by regarding one of the original state variables as a fictitious time. In spite of this, it is useful to think in terms of autonomous versus nonautonomous, since many properties of the dynamics are immediate from such considerations.

Illustration

Consider an earth-orbiting satellite moving under the combined gravitational influence of the bodies of the solar system. The many-body system consisting of satellite, planets, and other solar system objects can for most practical purposes be considered autonomous, since the influence of bodies outside of the immediate vicinity of the sun can be neglected. A further degree of simplification (and quite substantial at that) can be introduced by considering the internal motions of the solar system as independent of the motion of the satellite. In fact, due to the vanishingly small mass of the satellite relative to the other major solar system bodies, the laws of mechanics imply that the latter will behave largely as if the satellite were absent. The satellite then simply moves in a time-dependent gravitational field and its evolution equation becomes nonautonomous.

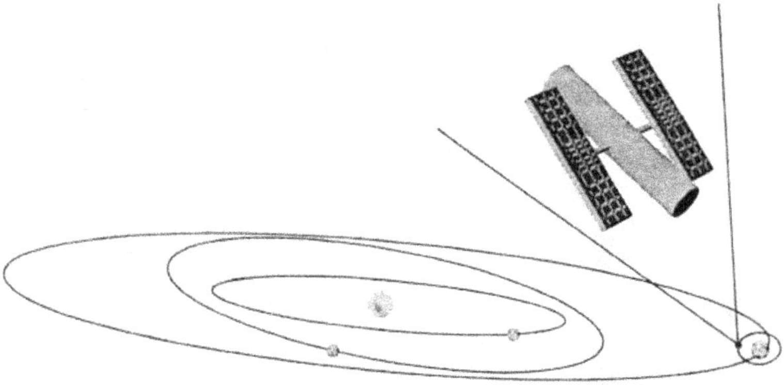

The orbital dynamics of most solar system objects experience a negligible influence from a geo-orbiting satellite.

Continuing in this manner, bar the presence of external deities, the entire classical universe should thus be described by autonomous evolution rules. The nonautonomous models that we choose to study are simply a consequence of choice of perspective. •

An autonomous system whose future is uniquely determined from its past and present is said to be **deterministic**. For example, a collection of classical particles, disconnected from the rest of the universe, would evolve in a deterministic fashion. Along similar lines, a Laplacian deity, knowing the initial values of all state variables describing the universe, would be able to predict its future development to the last decimal (pun intended). Classical thermodynamics, realizing the practical impossibility of simultaneous tracking of large numbers of particles, concerns itself with large-scale averages, rather than with individual spatial and temporal details. Consequently, its evolutionary rules are **statistical** and **stochastic**, rather than deterministic. The procedure replaces a very high-dimensional, deterministic system with a relatively low-dimensional, nondeterministic system. In this sense, the differentiation between determinism and nondeterminism seems a matter of one's ability to resolve details on different scales.

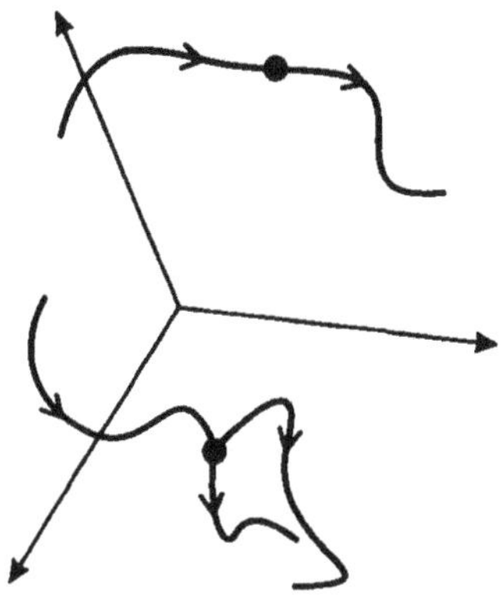

In a deterministic system, the situation in the lower part of the figure could not happen, i.e., the evolution of the state would be unique.

On the other hand, most modern physicists argue that the dynamics of subatomic particles is essentially stochastic and the apparent determinism observed in large-scale motions of macroscopic bodies is a result of spatial and temporal averaging. Thus, the appropriateness of using deterministic versus stochastic models should be judged on their usefulness in describing real-world observations. With a hint of what is to come, we suggest that even in low-dimensional deterministic systems, the observed behavior appears stochastic rather than predictable. This inherent stochasticity in some deterministic systems makes the issue even less clear cut.

Determinism is a one-sided feature of a dynamical system. While the future is, in principle, completely predictable based on the past and present, it is generally impossible to determine the past given future and present. Here again, the Laplacian universe requires a two-sided determinism, implying a possibility of determining the exact past of a current state of the universe. Suffice it to say that the discussion of the previous paragraph again applies and that the choice between one- or two-sided determinism should be made based on usefulness, rather than philosophy.

Dynamical systems come in essentially two types. **Discrete** dynamical systems are concerned only with the state at discrete moments in time. The evolution rule is usually an explicit expression of the form

$$\boldsymbol{x}^{i+1} = \boldsymbol{f}(i, \boldsymbol{x}^i),\ i \in \mathbb{Z} \tag{1.3}$$

for some function $\boldsymbol{f}$. Here, $\boldsymbol{x}^i$ refers to the state at (discrete) time i. The form of Eq. (1.3) implies that the future of the system is uniquely determined once an initial state, $\boldsymbol{x}^0$, is given; i.e., the system is deterministic. However, unless the mapping $\boldsymbol{f}$ is injective (one-to-one), the past is not unique and can quickly be forgotten. A convenient measure of the typical rate at which the past is forgotten is provided by the **autocorrelation function**

$$C(k) = \lim_{N\to\infty} \frac{1}{N} \sum_{i=1}^{N} \boldsymbol{x}^i \cdot \boldsymbol{x}^{i+k}. \tag{1.4}$$

For a typical orbit, the correlation function decreases with increasing k. The characteristic time required to erase the details of the past could, for example, be the value of k for which $C(k)$ has decreased by a predetermined amount.

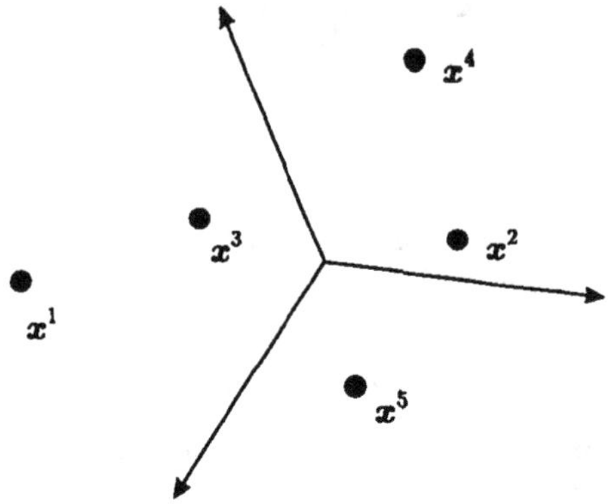

An orbit segment for a discrete dynamical system.

Illustration

An example is given by considering the doubling map:

$$x^{i+1} = 2x^i \bmod 1, \text{ where } x \in [0,1]. \tag{1.5}$$

At each successive step the current value of the single state variable, x^i, is doubled and translated back into the original interval. The evolution of the state is clearly determined by the choice of initial value. Note, however, that attempting to retrace the past of a current state meets with immediate difficulty, since the map is 2-1. Thus, for example, $x^0 = a$ and $x^0 = \frac{1}{2} + a$, $a < \frac{1}{2}$, are both mapped onto the same point. The complete dynamics of this dynamical system are most naturally illustrated by switching to a binary representation of state space, in which a state corresponds to a binary decimal. The doubling map simply shifts the

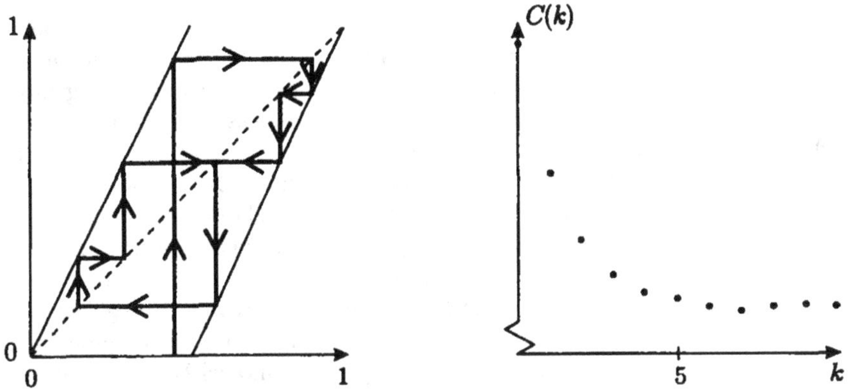

The left panel shows a typical periodic orbit. The thin lines illustrate the mapping $\boldsymbol{f}$ *and the dashed line, the identity mapping. The right panel shows the typical autocorrelation function for an orbit corresponding to an irrational initial point.*

decimal representation one step to the left and throws away the integer part. It follows that the asymptotic behavior of an orbit is determined by the succession of ones and zeros far downstream.

Clearly, any binary decimal integer with finitely nonzero decimals will eventually be mapped to the origin. Similarly, a rational number has an eventually repeating sequence of decimals and thus the orbit eventually ends up in a loop. Finally, a state with an irrational decimal representation has an orbit that never repeats and gets arbitrarily close to all points in the interval. •

For **continuous** dynamical systems, one is concerned with the value of the state at each moment in an open interval of time. The corresponding evolution rules are mostly ordinary (ODEs) or partial (PDEs) differential equations. This book concerns itself with the finite-dimensional case for which ODEs suffice. The typical evolution rule thus has the form

$$\dot{\boldsymbol{x}} = \boldsymbol{f}(t, \boldsymbol{x}),\ t \in \mathbb{R} \tag{1.6}$$

for some function $\boldsymbol{f}$ known as the **vector field**. Solving this equation yields the state $\boldsymbol{x}(t)$ as an explicit function of time. Thus, in order for an equation of the form (1.6) to describe a dynamical system, such solutions must exist.

Under rather weak smoothness conditions on $\boldsymbol{f}$, fundamental theorems on local **existence** and **uniqueness** of solutions imply that, at least for short intervals of time, the future and past of the system are uniquely determined from the present state at a given time. However, existence cannot be guaranteed for all times. While one could certainly imagine solutions that cease to exist in finite time as describing physical systems whose characteristics change dramatically, we will exclusively consider ODEs for which **global** solutions exist. In other words, there will be no limitations on the intervals of time over which the solutions can be extended. In the autonomous case this will also imply global uniqueness and hence two-sided determinism.

The analysis of a given dynamical system of the types described above consists of obtaining as complete a knowledge as possible about the different dynamical behaviors contained therein. A comprehensive numerical search through choosing a satisfactory sample of initial conditions is often the first step in such a process. Still, it is difficult to put complete faith in such an approach, since without any other information one cannot be certain that one's sampling is sufficiently fine. Not to mention the possibility of numerical errors interfering with the analysis. There are, in fact, a large number of properties discernible through analytic means, the simplest of which is the analysis of invariant sets to which we now turn.

1.1.2 Flows and invariant sets

We consider sufficiently smooth dynamical systems for which global existence and uniqueness of solutions hold. Given an initial state $\boldsymbol{x}(s)$ at time s, we speak of the point $\boldsymbol{x}(t)$ as the **image** of the initial point under the evolution of the dynamical system after time $t-s$. The invertible set map, $\boldsymbol{\phi}(t,s,\cdot)$ that describes this relationship is called the **flow** of the dynamical system. It is defined by considering its effect on singletons $\boldsymbol{x}(s)$:

$$\boldsymbol{\phi}(t,s,\boldsymbol{x}(s)) = \boldsymbol{x}(t),\ \forall t,s \in \mathbb{R}. \tag{1.7}$$

Clearly, $\boldsymbol{\phi}(t,s,\cdot) = \boldsymbol{\phi}^{-1}(s,t,\cdot)$ and $\boldsymbol{\phi}(s,s,\cdot)$ is the identity map. One can further

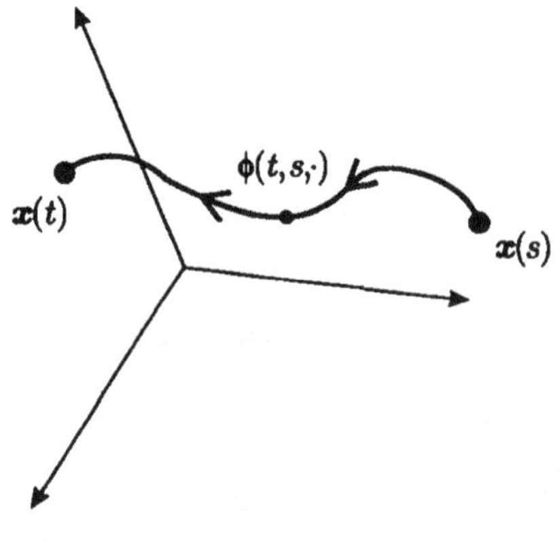

The flow maps states to states further along their orbits.

show that ϕ is as smooth as f. In our previous terminology, $\phi(t, s, x(s))$, $t \geq s$ maps $x(s)$ to its forward orbit, etc. Naturally, the flow terminology comes by analogy from fluid mechanics where the vector field is the velocity field of a fluid and the flow simply keeps track of the position of individual fluid particles.

Illustration

For a linear, continuous dynamical system

$$\dot{x} = A(t)x, \tag{1.8}$$

where $A(t)$ is piecewise continuous in time, the existence of a nonsingular, fundamental matrix $X(t)$ allows the general solution to be written

$$x(t) = X(t)X^{-1}(s)x(s) \tag{1.9}$$

or, in other words,

$$\phi(t, s, x) = X(t)X^{-1}(s)x. \tag{1.10}$$

We note that, by linearity, the image of a subspace of S under this flow is still a subspace.

In the autonomous case, $A(t) \equiv A$, the fundamental solution matrix is given by

$$X(t) = e^{tA} = \sum_{k=0}^{\infty} \frac{(tA)^k}{k!}. \tag{1.11}$$

Consequently, the flow is simply the matrix operator

$$\phi(t, s, \cdot) = e^{(t-s)A}, \tag{1.12}$$

i.e., $\phi(t, s, \cdot)$ depends only on the elapsed time. This is actually true for all autonomous dynamical systems, i.e.,

$$\phi(t, s, \cdot) = \phi(t - s, \cdot) \tag{1.13}$$

when $f(t, x) = f(x)$. •

A family of sets, $\mathcal{U}_t$, parameterized by $t \in \mathcal{I}$ is said to be ϕ-compatible if $\mathcal{U}_t = \phi(t, s, \mathcal{U}_s)$ for all $s, t \in \mathcal{I}$. Such a family will be denoted $\mathcal{U}_t = \mathcal{U}(t)$. In light of the fluid mechanics analogy mentioned above, a family of ϕ-compatible sets contains the same "particles" over the relevant time interval. Given a family, $\mathcal{U}(t)$, $t \in \mathbb{R}$, we say that a family $\mathcal{V}(t)$ is **attracted** to $\mathcal{U}(t)$ if

$$d\left(x(t), \mathcal{U}(t)\right) \to 0 \text{ as } t \to \infty \text{ for all } x(t) \in \mathcal{V}(t) \tag{1.14}$$

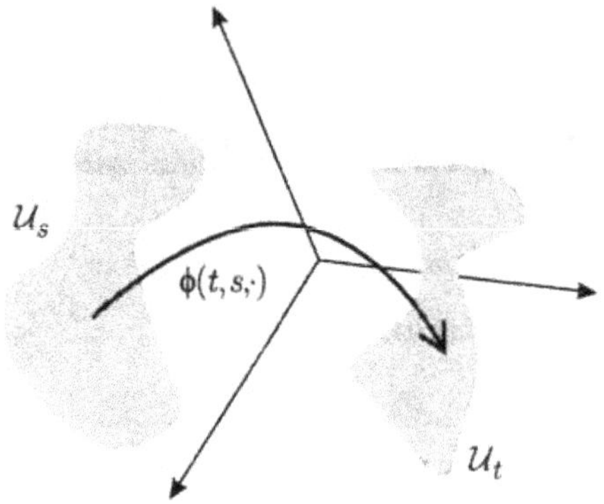

The sets $\mathcal{U}_s$ and $\mathcal{U}_t$ belong to a ϕ-compatible family.

where $d(x,\mathcal{U})$ is the distance between the point x and the set $\mathcal{U}$ given by the metric on state space. For simplicity we write $\mathcal{V}(t) \to_{\text{p.w.}} \mathcal{U}(t)$. The family $\mathcal{U}^s(t) = \bigcup \mathcal{V}(t)$ where the union is over all $\mathcal{V}(t)$ such that $\mathcal{V}(t) \to_{\text{p.w.}} \mathcal{U}(t)$ is called the **inset** of $\mathcal{U}(t)$. Similarly, we define the **outset** $\mathcal{U}^u(t)$ by considering sets attracted to $\mathcal{U}(t)$ as $t \to -\infty$.

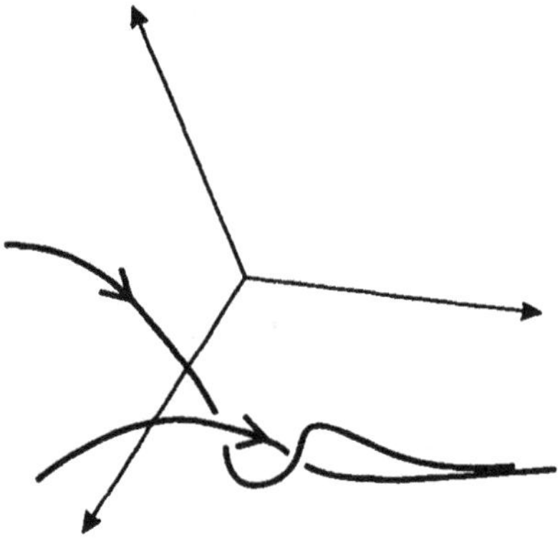

The two orbits attract each other as $t \to \infty$.

Of particular interest in the study of dynamical systems are sets that are **invariant** under the flow, i.e.; sets $\mathcal{U}$, such that

$$\phi(t,s,\mathcal{U}) \subseteq \mathcal{U}, \forall t,s \in \mathbb{R}. \tag{1.15}$$

In this context, we also speak of **positively** and **negatively** invariant sets as being sets satisfying the above inclusion for all $t \geq s$ and $t \leq s$, respectively. It is easy to show that invariant sets are actually **fixed sets**, i.e., that equality holds in Eq. (1.15), since $\mathcal{U} = \phi(t, s, \phi(s, t, \mathcal{U})) \subseteq \phi(t, s, \mathcal{U})$, $\forall t, s \in \mathbb{R}$.

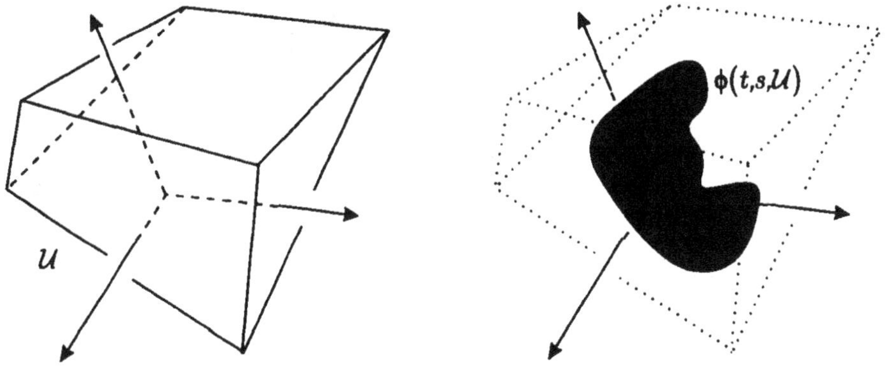

The set $\mathcal{U}$ is positively invariant if the image lies inside $\mathcal{U}$ for all $t \geq s$.

Evidently, for autonomous systems, the forward (backward) orbit of an initial state is positively (negatively) invariant, and the entire orbit is an invariant set. Similarly, collections of orbits satisfy these properties. For example, an inset is positively invariant. More interesting than arbitrary collections of orbits are invariant manifolds. Since manifolds allow the application of all the standard techniques of many-variable analysis, these latter types of invariant sets lend themselves more easily to geometric and perturbative techniques. We will return to these issues in later sections.

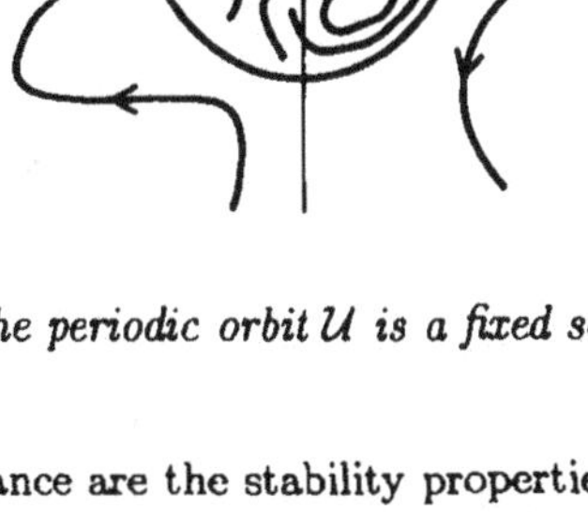

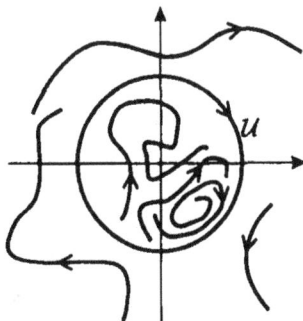

The periodic orbit $\mathcal{U}$ is a fixed set.

Of practical importance are the stability properties of invariant sets, i.e., the asymptotic, long-term behavior of nearby solutions with respect to their distance from

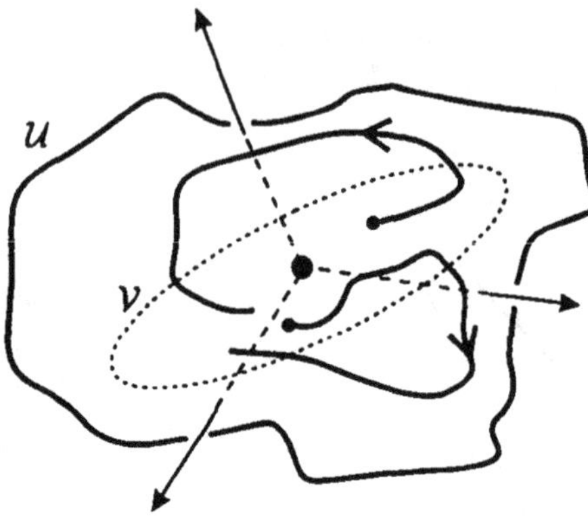

The invariant set at the origin is stable.

the set. For example, an invariant set is **stable** if, for any given neighborhood $\mathcal{U}$ of the set, there is a neighborhood $\mathcal{V} \in \mathcal{U}$ such that all orbits initially in $\mathcal{V}$ remain in $\mathcal{U}$ for all positive time, and **unstable** otherwise. We speak of **asymptotic stability** for positive (negative) time, if the orbits starting in $\mathcal{V}$ actually approach the set arbitrarily close as time goes to plus (minus) infinity. Closed sets that are asymptotically stable for positive time are called **attracting**. Similarly, asymptotically stable closed sets in negative time are **repelling**.

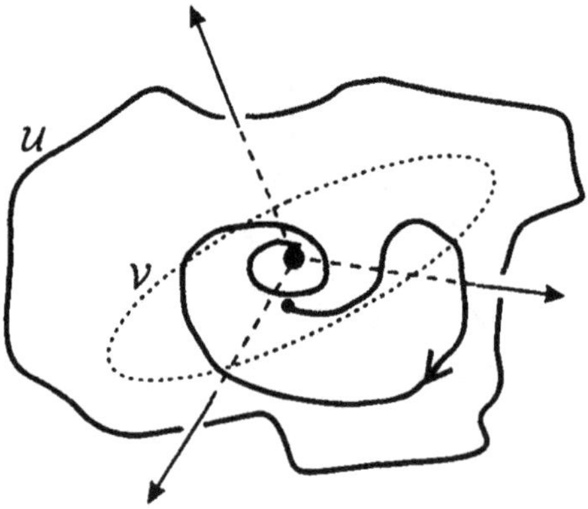

The origin is attracting.

Illustration

Consider the flow given by the dynamical system

$$\dot{r} = \frac{1}{10}|1 - r|, \dot{\theta} = 1 \qquad (1.16)$$

where (r, θ) are polar coordinates. The phase portrait of this dynamical system is shown below. It is immediately clear that the family $\mathcal{U}(t) = \{r = 1\} \times [0, 2\pi)$ is fixed under the flow. Its inset and outset consist of $\mathcal{U}(t)$ itself, together with all points inside and outside the unit circle, respectively. Thus, $\mathcal{U}(t)$ is unstable; however, it is not repelling.

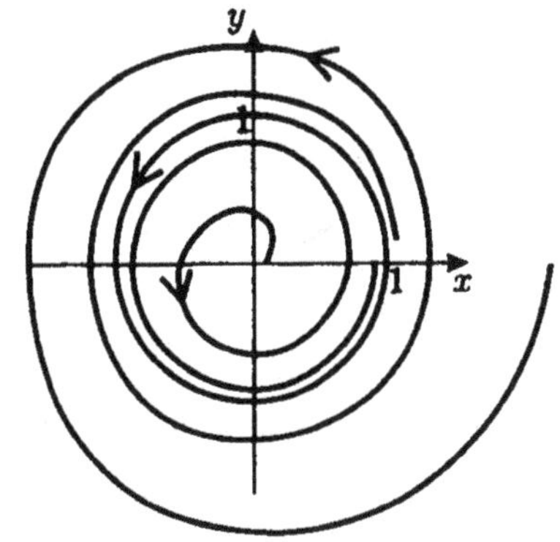

The phase portrait for the dynamical system in the illustration.

If we instead consider the invariant family $\mathcal{U}(t) = \{r \leq 1\} \times [0, 2\pi)$, then indeed any family of open sets containing $\mathcal{U}(t)$ is repelled from $\mathcal{U}(t)$. Thus, the outset is all of state space and the inset is $\mathcal{U}(t)$ itself. Consequently, $\mathcal{U}(t)$ is a repelling set. •

1.1.3 Stationary points and their stability

Particularly simple examples of orbits are **stationary points**, or **equilibria** and **fixed points** for continuous and discrete systems, respectively. In the case of continuous dynamical systems, these can be found by equating the vector field at the equilibrium to zero and solving for the equilibrium coordinates. Similarly, in the discrete case, substituting the fixed point for all occurrences of state vectors in the evolution rule and solving yields the fixed point's coordinates. Finding the stationary points of a dynamical system is usually one of the first steps taken in studying its dynamics. On the one hand, this is because they are theoretically straightforward to find. (We note that in most cases numerical methods are necessary and exhaustive searches for all stationary points can be practically impossible.) On the other hand, it turns out that their presence and their properties yield much information about the dynamical system, at least for initial states nearby. For the time being, we shall restrict attention to autonomous systems.

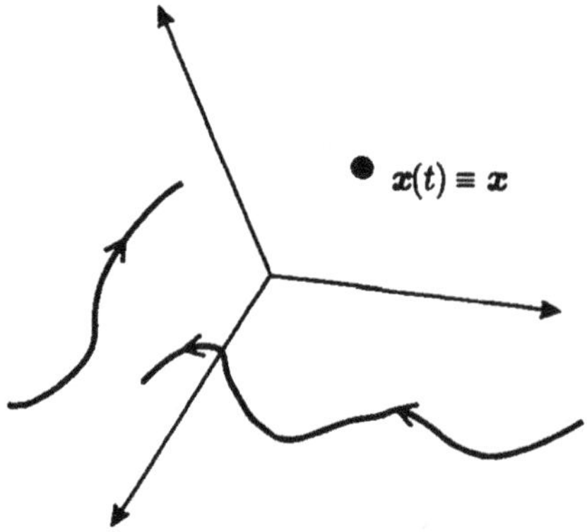

x is a stationary point of the flow.

For linear dynamical systems,

$$\dot{x} = Ax, \tag{1.17}$$

there is an equilibrium at the origin. Its stability properties are found from an analysis of the eigenvalues of the corresponding matrix, A. The general solution can be written

$$x = \sum_{k=1}^{m} \left(a_0^k + a_1^k t + \ldots + a_{\mu_k - 1}^k t^{\mu_k - 1} \right) e^{\lambda_k t}, \tag{1.18}$$

where λ_k, μ_k denote A's k-th eigenvalue and its multiplicity. Further, $n = \mu_1 + \ldots \mu_m$ is the dimension of the dynamical system. The coefficients a_j^k are constant vectors supplied by the initial conditions and the form of A. From this expression, we see that if $\Re(\lambda_k) < 0$ (> 0) $\forall k$, then $x \to \mathbf{0}$, as $t \to \infty$ $(t \to -\infty)$, i.e., the origin is asymptotically stable in forward (backward) time.

Similarly, if $\Re(\lambda_k) > 0$ (< 0) for some k, then the origin is unstable in forward (backward) time. If an eigenvalue lies on the imaginary axis, stability is, in general, only possible if its multiplicity is one, since otherwise any polynomial coefficients would grow without bounds. Similarly, stability for all times can only be achieved if all eigenvalues lie on the imaginary axis. The origin will not, however, be attracting, since nearby solutions will stay bounded away from it.

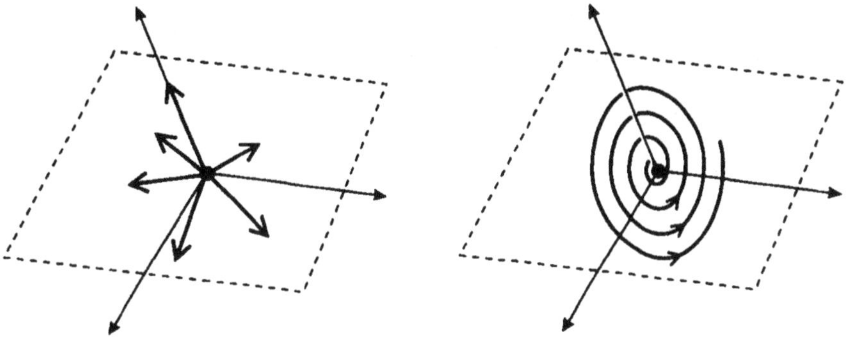

Examples of the flow around stationary points in linear systems.

Since Eq. (1.18) is a general solution, it follows that the conclusions are global. In other words, an attracting equilibrium is actually globally attracting, and so on. We refer to attracting equilibria as **sinks**, and their reverse time equivalents as **sources**. Further, equilibria with eigenvalues in both half planes but away from the imaginary axis will be denoted **saddles** and prove essential to our discussions to come. These various types of stationary points are all known as **hyperbolic** points. Nonhyperbolic, or **degenerate**, points will not be discussed in this text. They often appear as signatures of bifurcations, or indicate the presence of particular symmetries.

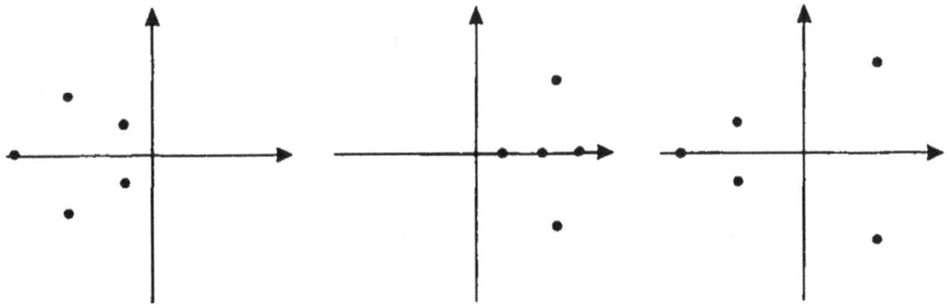

Examples of eigenvalue distributions for sinks, sources, and saddles, respectively.

Illustration

For discrete, linear dynamical systems

$$\boldsymbol{x}^{i+1} = \boldsymbol{A}\boldsymbol{x}^{i} \tag{1.19}$$

we essentially replace $e^{\lambda_k t} \to \lambda_k^i$ *and* $t \to i$ *in Eq. (1.18). The above division of the complex plane into a left and right half-plane now becomes the inside and outside of the unit circle. Similarly, the separating imaginary axis becomes the unit circle itself. Thus, for example, asymptotic stability of the fixed point at the origin is guaranteed if all eigenvalues lie within the unit circle, etc.* •

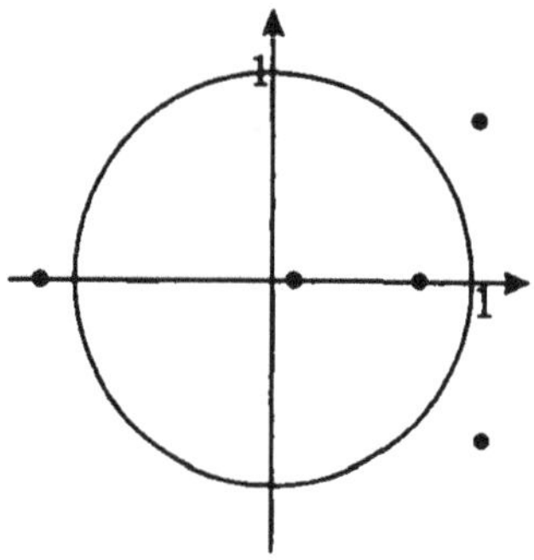

The eigenvalues of a linear discrete system with a saddle point at the origin.

It is natural to consider how these stability notions carry over to a nonlinear system with vector field, $\boldsymbol{f}(\boldsymbol{x})$. A very useful result is that of *Hartman-Grobman*. It states that under certain conditions stationary points of nonlinear systems retain their stability properties from a local analysis. By a local analysis, one means Taylor expanding the vector field of the nonlinear dynamical system near the stationary point and throwing away all but the linear terms. In particular, if the stationary point is hyperbolic for the resulting linear system, then the Hartman-Grobman results apply. Consequently, a linear sink becomes a nonlinear sink, a linear saddle a nonlinear saddle, and so on. However, it is no longer true that a sink is automatically globally attracting, since nonlinear effects become increasingly important as one moves away from the stationary point. The Hartman-Grobman theorem is nonconclusive regarding degenerate equilibria, for which a higher-order analysis is usually necessary.

Illustration

For time-periodic continuous dynamical systems it is easy to see that

$$\boldsymbol{\phi}(s+mT,s,\boldsymbol{x})=\boldsymbol{\phi}(s+T,s,\boldsymbol{\phi}(s+(m-1)T,s,\boldsymbol{x})) \tag{1.20}$$

where T is the period of the vector field, s is an arbitrarily chosen reference time, and m is an arbitrary integer. It is convenient to associate to the

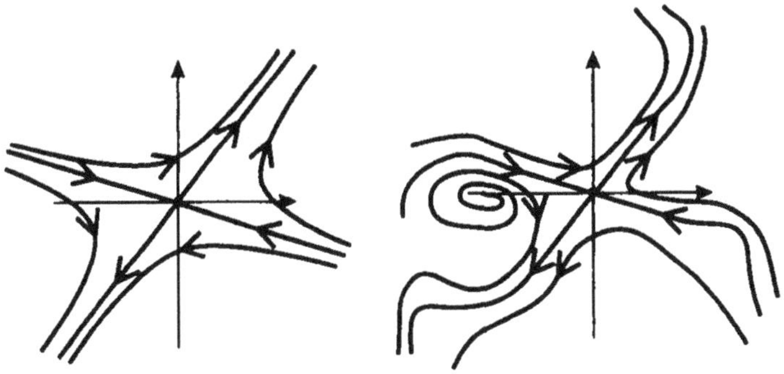

The Hartman-Grobman result implies that the linear analysis (left panel) applies locally in the nonlinear case (right panel).

flow the **period map**

$$\boldsymbol{f}_{T,s}(\boldsymbol{x}) = \boldsymbol{\phi}(s+T, s, \boldsymbol{x}) \tag{1.21}$$

so that

$$\boldsymbol{f}^m_{T,s}(\boldsymbol{x}) = \boldsymbol{\phi}(s+mT, s, \boldsymbol{x}). \tag{1.22}$$

Moreover,

$$\boldsymbol{\phi}(s+T, s, \boldsymbol{\phi}(s,0,\boldsymbol{x})) = \boldsymbol{\phi}(s,0,\boldsymbol{\phi}(T,0,\boldsymbol{x})) \tag{1.23}$$

implies

$$\boldsymbol{f}_{T,s} \circ \boldsymbol{\phi}(s,0) = \boldsymbol{\phi}(s,0) \circ \boldsymbol{f}_{T,0}, \tag{1.24}$$

i.e., the homeomorphism $\boldsymbol{\phi}(s,0)$ takes orbits of $\boldsymbol{f}_{T,0}$ onto those of $\boldsymbol{f}_{T,s}$ preserving the direction of increasing time. The behavior of the discrete dynamical system governed by $\boldsymbol{f}_{T,s}$ is thus closely related to that of $\boldsymbol{f}_{T,0}$. Consequently, it suffices to study $\boldsymbol{f}_{T,0}$ and we will henceforth omit the 0 subscript.

We consider an mT-periodic orbit, $\boldsymbol{x}(t)$, of the original flow such that $\boldsymbol{x}(0) = \boldsymbol{x}_0$. Clearly, $\boldsymbol{x}_0$ is a fixed point of $\boldsymbol{f}^m_T$, i.e. $\boldsymbol{f}^m_T(\boldsymbol{x}_0) = \boldsymbol{x}_0$. If all the eigenvalues of the linearization of the period-m map at $\boldsymbol{x}_0$, $(\boldsymbol{D}\boldsymbol{f}^m_T)(\boldsymbol{x}_0)$, lie inside the unit circle, then the original periodic orbit is locally asymptotically stable. Similar conclusions follow for other types of hyperbolic fixed points of $\boldsymbol{f}^m_T$. •

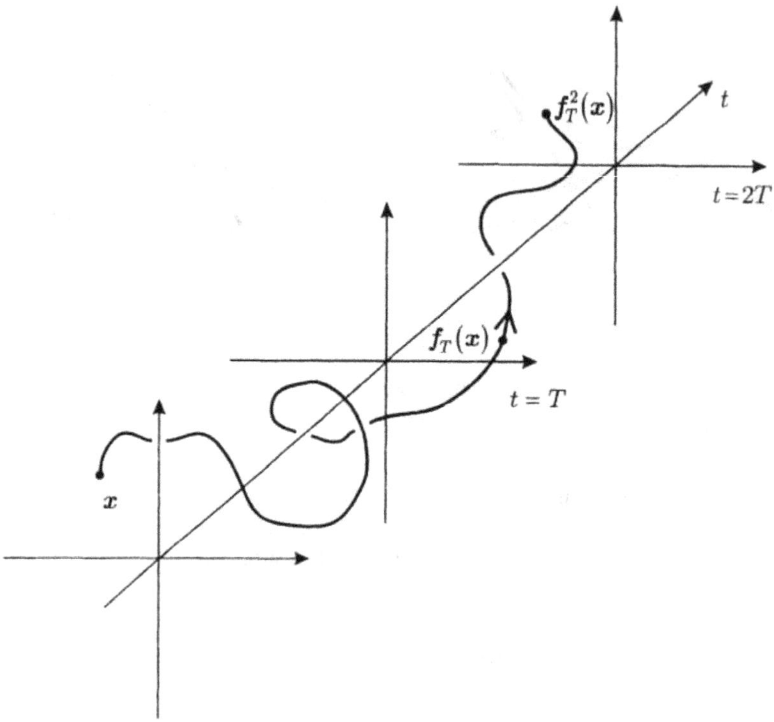

The period map.

1.1.4 The stable and unstable manifolds

In the previous section we saw that the linear hyperbolicity of a stationary point ensured that its stability properties carried over to the nonlinear flow. In this section we shall see that, in fact, additional global information follows from this hyperbolicity. The objects introduced in this section will be of fundamental importance throughout the text and we will therefore attempt to be especially careful at this point. Of particular importance will be estimates of contraction and expansion rates for vectors in some complementary subspaces.

We again consider a linear, autonomous system

$$\dot{\boldsymbol{x}} = \boldsymbol{A}\boldsymbol{x} \tag{1.25}$$

with an equilibrium at the origin. From linear algebra we learn that state space can be written as a direct sum of two complementary invariant subspaces, $\mathcal{E}^{s,u}$, such that an arbitrary state can be uniquely decomposed as

$$\boldsymbol{x} = \boldsymbol{x}_s + \boldsymbol{x}_u, \ \boldsymbol{x}_s \in \mathcal{E}^s, \boldsymbol{x}_u \in \mathcal{E}^u \tag{1.26}$$

and, furthermore,

$$\phi(t, \boldsymbol{x}_s) \to \mathbf{0} \text{ as } t \to \infty \tag{1.27}$$

and similarly for $\boldsymbol{x}_u$ as $t \to -\infty$. $\mathcal{E}^s$ and $\mathcal{E}^u$ are known as the **stable** and **unstable subspaces**, respectively. From the general solution (1.18), it follows that the component in the stable subspace decays exponentially toward the equilibrium in forward time. Similarly, states in $\mathcal{E}^u$ are repelled at an exponential rate from the equilibrium. The dimensionality of the stable (unstable) subspace equals the number (including multiplicity) of eigenvalues in the left (right) half plane. For sinks (sources), the unstable (stable) subspace is consequently empty, while for saddles the subspaces are each at least one-dimensional.

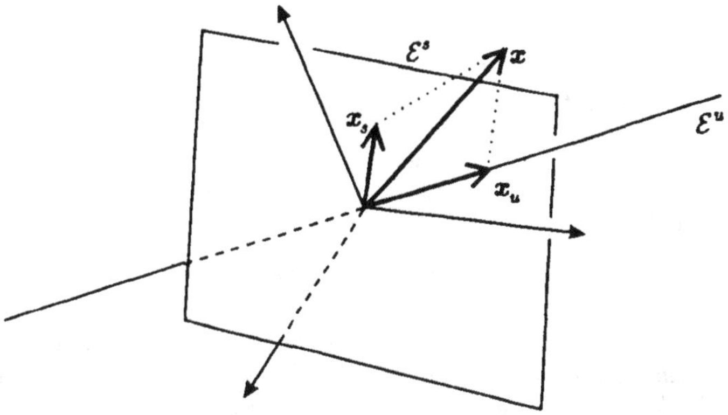

The decomposition of an arbitrary vector into its stable and unstable components along the stable and unstable subspaces, respectively.

Since the subspaces $\mathcal{E}^{s,u}$ are complementary, there exists a projection operator, $\boldsymbol{P}$, such that

$$\text{Range}(\boldsymbol{P}) = \mathcal{E}^s, \ \text{Ker}(\boldsymbol{P}) = \mathcal{E}^u, \tag{1.28}$$

and $\boldsymbol{P}^2 = \boldsymbol{P}$. Let

$$K = |\boldsymbol{P}| = \sup_{|\boldsymbol{x}|=1} |\boldsymbol{P}\boldsymbol{x}| \tag{1.29}$$

denote the norm of $\boldsymbol{P}$, from which we get $|\boldsymbol{I} - \boldsymbol{P}| \leq 1 + K$. Then

$$|\phi(t) \circ \boldsymbol{P}| = \sup_{|\boldsymbol{x}|=1} |\phi(t, \boldsymbol{P}\boldsymbol{x})| \leq K' e^{-\lambda t}, t \geq 0 \tag{1.30}$$

and similarly

$$|\phi(t) \circ (\boldsymbol{I} - \boldsymbol{P})| = \sup_{|\boldsymbol{x}|=1} |\phi(t, \boldsymbol{x} - \boldsymbol{P}\boldsymbol{x})| \leq K' e^{\lambda t}, t \leq 0 \tag{1.31}$$

for $0 < \lambda \leq \min_k |\Re(\lambda_k)|$, and some $K' \geq \max\{K, 1+K\}$. In other words, the projection of an arbitrary initial state onto the stable subspace $\mathcal{E}^s$ decays in forward time at a rate bounded below by the absolute value of the real part of the least negative eigenvalue. Similarly, the complementary projection decays in backward time at a rate bounded below by the real part of the least positive eigenvalue.

The inequalities Eq. (1.30-1.31) are characteristic of the hyperbolicity of the fixed point. Below we derive generalizations of these inequalities to systems with similar properties. In Chapter 3 these inequalities will in themselves be taken as characterizing properties of certain linear dynamical systems. This will then be used to discuss the presence of chaotic, stochastic behavior in otherwise deterministic systems.

Illustration

We consider a time-periodic linear system

$$\dot{\boldsymbol{x}} = \boldsymbol{A}(t)\boldsymbol{x} \tag{1.32}$$

for which the period map is

$$\boldsymbol{f}_T(\boldsymbol{x}) = \boldsymbol{\phi}(T, 0, \boldsymbol{x}) = \boldsymbol{X}(T)\boldsymbol{X}^{-1}(0)\boldsymbol{x}. \tag{1.33}$$

Thus, $\boldsymbol{f}_T$ is linear and the stability of the origin is determined by the locus of the eigenvalues of $\boldsymbol{X}(T)\boldsymbol{X}^{-1}(0)$. We assume that these all lie off the unit circle. As in the continuous case, there then exists an invariant decomposition $\mathcal{S} = \mathcal{E}^s \oplus \mathcal{E}^u$, where the dimension of $\mathcal{E}^s$ equals the number of eigenvalues (including multiplicity) inside the unit circle and similarly for $\mathcal{E}^u$.

We now consider the family of subspaces $\mathcal{E}^{s,u}(t)$ such that $\mathcal{E}^{s,u}(0) = \mathcal{E}^{s,u}$. In fact, $\mathcal{S} = \mathcal{E}^s(s) \oplus \mathcal{E}^u(s)$ is the decomposition corresponding to the period map based at $t = s$: $\boldsymbol{f}_{T,s}$. If we denote the projection operator corresponding to $\mathcal{E}^{s,u}(0)$ by $\boldsymbol{P}(0)$, then the periodic function

$$\boldsymbol{P}(t) = \boldsymbol{X}(t)\boldsymbol{X}^{-1}(0)\boldsymbol{P}(0)\boldsymbol{X}(0)\boldsymbol{X}^{-1}(t) \tag{1.34}$$

is the corresponding projection at time t and indeed

$$\boldsymbol{\phi}(t, s, \boldsymbol{P}(s)\boldsymbol{x}) = \boldsymbol{P}(t)\boldsymbol{\phi}(t, s, \boldsymbol{x}). \tag{1.35}$$

Let $K = |\boldsymbol{P}(0)|$ and

$$c_1 = \sup_{t\in[0,T]} \left|\boldsymbol{X}(t)\boldsymbol{X}^{-1}(0)\right| \tag{1.36}$$

$$c_2 = \sup_{t\in[0,T]} \left|\boldsymbol{X}(0)\boldsymbol{X}^{-1}(t)\right|. \tag{1.37}$$

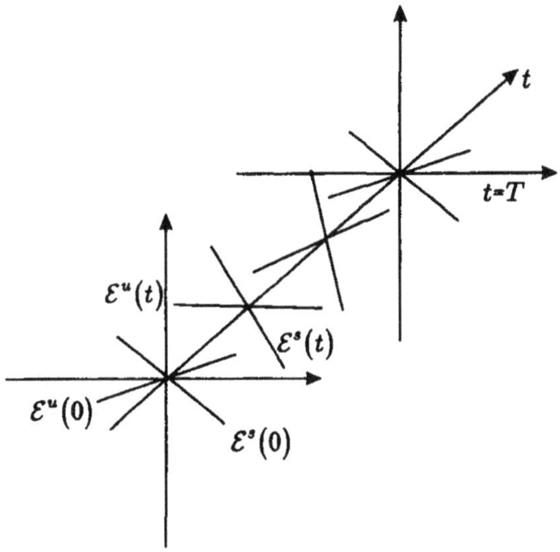

The family of subspaces $\mathcal{E}^{s,u}(t)$ considered in the illustration.

Then, as in the autonomous case, we have

$$|\boldsymbol{f}_T^m \circ \boldsymbol{P}(0)| \leq K'\lambda^m, m \geq 0 \tag{1.38}$$

and

$$|\boldsymbol{f}_T^m \circ (\boldsymbol{I} - \boldsymbol{P}(0))| \leq K'\lambda^{-m}, m \geq 0, \tag{1.39}$$

where $K' \geq \max\{K, 1+K\}$ and $0 < \lambda < 1$ such that λ is greater than all eigenvalues of $\boldsymbol{f}_T$ inside the unit circle and λ^{-1} is smaller than all those outside.

Moreover,

$$\left|\boldsymbol{f}_{T,s}^m \circ \boldsymbol{P}(s)\right| \leq K''\lambda^m, m \geq 0 \tag{1.40}$$

and

$$\left|\boldsymbol{f}_{T,s}^m \circ (\boldsymbol{I} - \boldsymbol{P}(s))\right| \leq K''\lambda^{-m}, m \leq 0 \tag{1.41}$$

where $K'' \geq \max\{1, K'c_1c_2\}$. Finally, if $s - nT = \tilde{s} \in [0, T]$ and $t - mT = \tilde{t} \in [0, T]$, $m \geq n$, then

$$\begin{aligned} \boldsymbol{\phi}(t, s, \boldsymbol{P}(s)\boldsymbol{x}) &= \boldsymbol{\phi}(t - nT, \tilde{s}, \boldsymbol{P}(\tilde{s})\boldsymbol{x}) \\ &= \boldsymbol{f}_{T,\tilde{t}}^{m-n} \circ \boldsymbol{\phi}(\tilde{t}, \tilde{s}, \boldsymbol{P}(\tilde{s})\boldsymbol{x}) = \boldsymbol{f}_{T,\tilde{t}}^{m-n} \circ \boldsymbol{P}(\tilde{t})\boldsymbol{\phi}(\tilde{t}, \tilde{s}, \boldsymbol{x}) \end{aligned} \tag{1.42}$$

where the first equality follows from the periodicity of the flow. Using the above relations and bounds we now obtain

$$|\boldsymbol{\phi}(t, s) \circ \boldsymbol{P}(s)| \leq K'''\lambda^{t-s}, t \geq s \tag{1.43}$$

and

$$|\phi(t,s)\circ(\boldsymbol{I}-\boldsymbol{P}(s))| \leq K'''\lambda^{s-t}, t \leq s \tag{1.44}$$

for some $K''' \geq 1$ (cf. Eq. (1.30-1.31). •

If we consider the family $\mathcal{U}(t) = \{\mathbf{0}\}$, it follows that $\mathcal{U}^{s,u}(t) = \mathcal{E}^{s,u}$. More generally, for an autonomous nonlinear system

$$\dot{\boldsymbol{x}} = \boldsymbol{f}(\boldsymbol{x}) \tag{1.45}$$

with a nondegenerate equilibrium at $\boldsymbol{x}_0$, the linearization at $\boldsymbol{x}_0$ admits an invariant subspace decomposition $\mathcal{E}^{s,u}(\boldsymbol{x}_0)$, such that $\{\boldsymbol{x}_0\}^{s,u} = \mathcal{E}^{s,u}(\boldsymbol{x}_0)$ for the linearized flow.

We now consider a neighborhood $\mathcal{V}$ of $\boldsymbol{x}_0$ and define the local inset of $\{\boldsymbol{x}_0\}$ for the nonlinear flow as

$$\mathcal{W}^s_{loc}(\boldsymbol{x}_0) = \{\boldsymbol{x} \in \mathcal{V} \mid \phi(t,\boldsymbol{x}) \to \boldsymbol{x}_0,\ t \to \infty \text{ and } \phi(t,\boldsymbol{x}) \in \mathcal{V}, \forall t \geq 0\} \tag{1.46}$$

and similarly for the local outset. From the Hartman-Grobman theorem we know that, in the hyperbolic case, the flow in the vicinity of the equilibrium is essentially that of the linearization. The question is then what relation there is between the inset/outset of the linearized flow and those of the nonlinear flow. The answer is provided by the *stable manifold theorem* from which it follows that $\mathcal{W}^{s,u}_{loc}(\boldsymbol{x}_0)$ are manifolds with a quadratic tangency to $\mathcal{E}^{s,u}(\boldsymbol{x}_0)$, at $\boldsymbol{x}_0$, and such that the dimensions of $\mathcal{W}^{s,u}_{loc}(\boldsymbol{x}_0)$ equal those of $\mathcal{E}^{s,u}(\boldsymbol{x}_0)$, respectively.

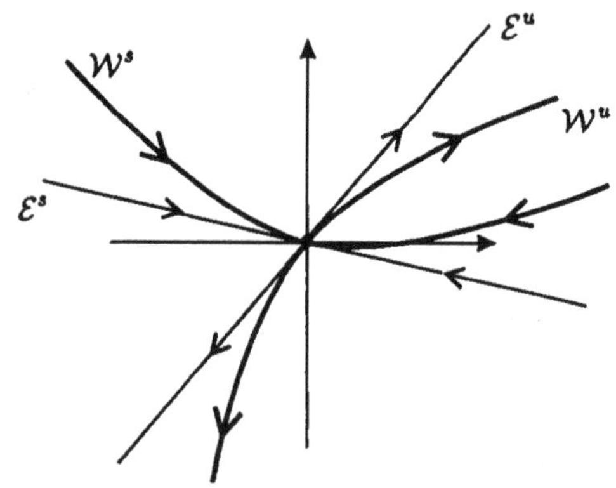

The stable manifold theorem.

By applying the flow to the local manifolds, we obtain global invariant manifolds, $\mathcal{W}^{s,u}(\boldsymbol{x}_0)$. These are known as the **stable** and **unstable manifolds**. Moreover, the contraction (expansion) on $\mathcal{E}^{s,u}(\boldsymbol{x}_0)$ is carried over to $\mathcal{W}^{s,u}(\boldsymbol{x}_0)$. Let $T_z\mathcal{W}^s(\boldsymbol{x}_0)$ denote the tangent space to the stable manifold at $\boldsymbol{z}$. There are then constants $K \geq 1$, $\lambda > 0$ such that for a given $\boldsymbol{y} \in \mathcal{W}^s(\boldsymbol{x}_0)$, and any $s \geq 0$

$$|\boldsymbol{\phi}_{\boldsymbol{x}}(t, \boldsymbol{z})\boldsymbol{\eta}| \leq Ke^{-\lambda t}|\boldsymbol{\eta}|, \; t \geq s \text{ for } \boldsymbol{\eta} \in T_z\mathcal{W}^s(\boldsymbol{x}_0) \tag{1.47}$$

where $\boldsymbol{z} = \boldsymbol{\phi}(s, \boldsymbol{y})$. In other words, tangent vectors to the stable manifold are exponentially contracted under the flow (recall that tangent vectors are mapped by the differential $\boldsymbol{\phi}_{\boldsymbol{x}}$). A similar expression holds for the unstable manifold for $t \leq s \leq 0$.

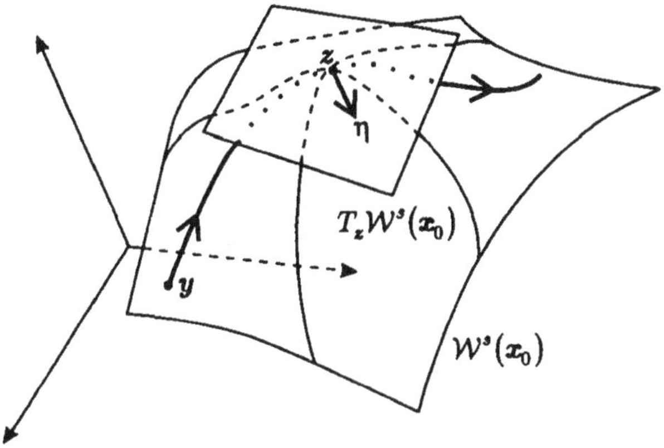

The global features of the stable manifold and its tangent space.

Using Eq. (1.47) and linearizing around the orbit $\boldsymbol{\phi}(s, \boldsymbol{y})$ on the stable manifold, we now wish to obtain estimates similar to Eq. (1.30-1.31). Consider the dynamical system whose flow is given by

$$\tilde{\boldsymbol{\phi}}(t, s, \boldsymbol{\eta}) = \boldsymbol{\phi}_{\boldsymbol{x}}\left(t - s, \boldsymbol{\phi}(s, \boldsymbol{y})\right)\boldsymbol{\eta}. \tag{1.48}$$

Note that $\boldsymbol{\phi}_{\boldsymbol{x}}(0, \boldsymbol{y}) = \boldsymbol{I}$, since $\boldsymbol{\phi}(0, \cdot)$ is the identity. The corresponding vector field is

$$\tilde{\boldsymbol{f}}(t, \boldsymbol{x}) = \boldsymbol{f}_{\boldsymbol{x}}(\boldsymbol{\phi}(t, \boldsymbol{y}))\boldsymbol{x}. \tag{1.49}$$

The dynamical system afforded by this vector field is known as the **variational equation** about the orbit $\boldsymbol{\phi}(s, \boldsymbol{y})$. Now consider the $\tilde{\boldsymbol{\phi}}$-compatible family of subspaces $\mathcal{E}(t)$, $t \geq 0$, such that $\mathcal{E}(0) = T_y\mathcal{W}^s(\boldsymbol{x}_0)$, and any projection $\boldsymbol{P}(t)$ onto $\mathcal{E}(t)$. Indeed, one sees that

$$\mathcal{E}(t) = T_{\boldsymbol{\phi}(t,\boldsymbol{y})}\mathcal{W}^s(\boldsymbol{x}_0). \tag{1.50}$$

It follows from Eqs. (1.47,1.48,1.50) that

$$|\tilde{\boldsymbol{\phi}}(t, s) \circ \boldsymbol{P}(s)| \leq K'e^{-\lambda(t-s)}, t \geq s \geq 0 \tag{1.51}$$

for some $K' \geq 1$. From the stable manifold theorem it follows that any complementary subspace would eventually become aligned with the unstable subspace. A local analysis together with a compactness argument then yields

$$|\tilde{\phi}(t,s) \circ (\boldsymbol{I} - \boldsymbol{P}(s))| \leq K'' e^{-\lambda(s-t)}, s \geq t \geq 0$$

for some $K'' \geq 1$ (cf. Eqs. (1.30-1.31, 1.43-1.44)). Similar considerations would apply if $\boldsymbol{y} \in \mathcal{W}^u(\boldsymbol{x}_0)$.

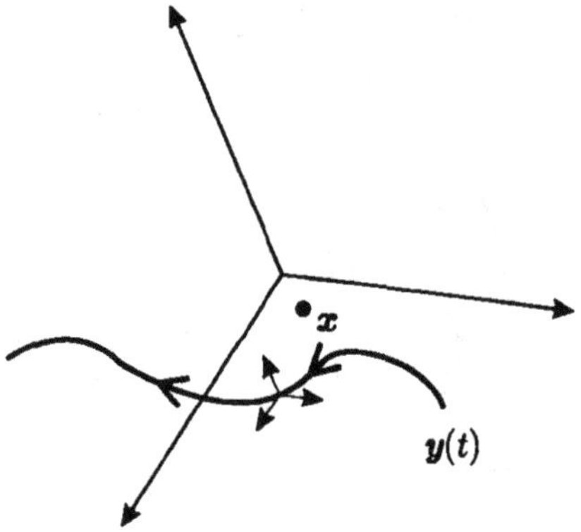

The variational equation describes local dynamics about a given orbit $\boldsymbol{y}(t)$ *of the original flow.*

Illustration

We repeat the above discussion for a hyperbolic periodic solution of period T to the dynamical system

$$\dot{\boldsymbol{x}} = \boldsymbol{f}(t, \boldsymbol{x}) \tag{1.52}$$

where $\boldsymbol{f}(t+T, \boldsymbol{x}) = \boldsymbol{f}(t, \boldsymbol{x})$. Such a periodic solution would appear as a fixed point, $\boldsymbol{x}_0$, of the period map $\boldsymbol{f}_T$. As in the previous illustration, it follows that there exists an invariant subspace decomposition $\mathcal{E}^{s,u}(\boldsymbol{x}_0)$ under the flow of $\boldsymbol{Df}_T(\boldsymbol{x}_0)$. The stable manifold theorem then implies the existence of global, invariant stable and unstable manifolds $\mathcal{W}^{s,u}(\boldsymbol{x}_0)$, tangent to $\mathcal{E}^{s,u}(\boldsymbol{x}_0)$ at $\boldsymbol{x}_0$. Moreover, for a point $\boldsymbol{y} \in \mathcal{W}^s(\boldsymbol{x}_0)$ there are constants $K \geq 1$, $0 < \lambda < 1$ such that

$$|\boldsymbol{Df}^m_{T,s}(\boldsymbol{z})\eta| \leq K\lambda^m|\eta|, \; m \geq 0 \text{ for } \eta \in T_z\mathcal{W}^s(\boldsymbol{x}_0) \tag{1.53}$$

for all $\boldsymbol{z} = \boldsymbol{\phi}(s, 0, \boldsymbol{y})$, where $\boldsymbol{\phi}(t, s, \cdot)$ is the flow corresponding to $\boldsymbol{f}(t, \boldsymbol{x})$. Applying this flow to $\mathcal{W}^{s,u}(\boldsymbol{x}_0)$ yields invariant manifolds that are the

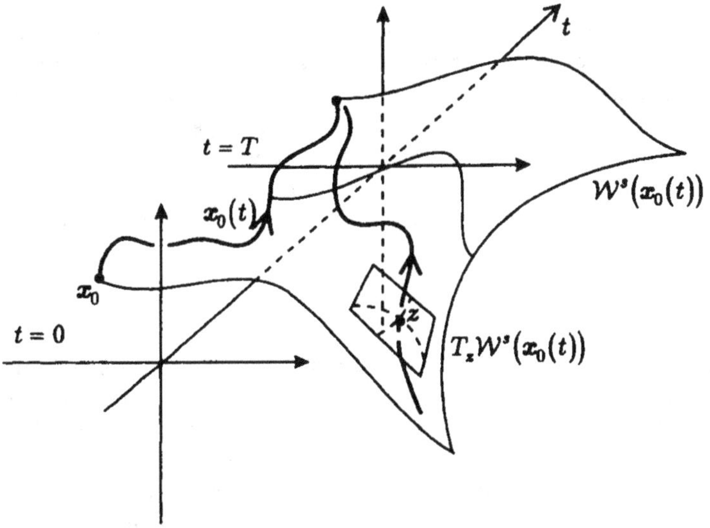

The stable manifold of a periodic orbit and its associated tangent space. Note that we are only concerned with tangent vectors parallel to the $t = 0$ plane.

inset and outsets of the family $\mathcal{U}(t) = \{\boldsymbol{\phi}(t, 0, \boldsymbol{x}_0)\}$. These are known as the stable and unstable manifolds of the periodic orbit.

Now define the flow

$$\tilde{\boldsymbol{\phi}}(t, s, \boldsymbol{\eta}) = \boldsymbol{\phi}_{\boldsymbol{x}}(t, s, \boldsymbol{\phi}(s, 0, \boldsymbol{y}))\boldsymbol{\eta} \tag{1.54}$$

corresponding to the vector field

$$\tilde{\boldsymbol{f}}(t, \boldsymbol{x}) = \boldsymbol{f}_{\boldsymbol{x}}(t, \boldsymbol{\phi}(t, \boldsymbol{y}))\boldsymbol{x}. \tag{1.55}$$

Furthermore, we consider the $\tilde{\boldsymbol{\phi}}$-compatible family

$$\mathcal{E}(t) = T_{\boldsymbol{\phi}(t,0,\boldsymbol{y})}\mathcal{W}^s(\boldsymbol{x}_0) \tag{1.56}$$

with a corresponding projection $\boldsymbol{P}(t)$. Using the periodicity of the original flow and the notation of the previous illustration, one finds

$$\boldsymbol{\phi}_{\boldsymbol{x}}\left(t, s, \boldsymbol{\phi}(s, 0, \boldsymbol{y})\right) \boldsymbol{P}(s)\boldsymbol{\eta}$$
$$= \boldsymbol{D}\boldsymbol{f}^{m-n}_{T,\tilde{t}}\left(\boldsymbol{\phi}\left(\tilde{t}, 0, \boldsymbol{f}^{n}_{T,s}(\boldsymbol{y})\right)\right) \boldsymbol{P}(\tilde{t})\boldsymbol{\phi}_{\boldsymbol{x}}(\tilde{t}, \tilde{s}, \boldsymbol{\phi}(\tilde{s}, 0, \boldsymbol{f}^{n}_{T,s}(\boldsymbol{y})))\boldsymbol{\eta}. \tag{1.57}$$

The uniform boundedness of the last operator on the right-hand side on $[0, T]$ together with Eq. (1.53) finally yields

$$|\tilde{\boldsymbol{\phi}}(t, s) \circ \boldsymbol{P}(s)| \le K'\lambda^{t-s},\ t \ge s \ge 0 \tag{1.58}$$

for some $K' \ge 1$. •

The existence of stable and unstable manifolds in a general flow has dramatic consequences on its dynamics. In particular, as we shall see in Chapter 3, this is the case when the manifolds intersect. The estimates for the contraction and expansion rates for vectors in complementary subspaces of state space, which we have obtained in this section, will be further generalized in Chapter 3. The availability of such estimates will turn out to be of fundamental importance and allow rather dramatic conclusions to be drawn about the global behavior of a large class of dynamical systems.

1.1.5 Conserved quantities

It is often useful to define real-valued functions with state space as domain of definition. Such functions are usually physically motivated, for example the kinetic energy of a system of rigid bodies. To avoid confusion with vector fields, such functions will be denoted using capital letters. Consider the smooth function $\boldsymbol{F} : \mathcal{S} \to \mathbb{R}^k$. We can imagine drawing a curve in $\mathbb{R}^k$ corresponding to a given orbit in $\mathcal{S}$. Of particular interest are functions for which the resulting curves are single points, regardless of the initial state. In other words, functions for which

$$\boldsymbol{F}(\boldsymbol{\phi}(t, s, \boldsymbol{x})) = \boldsymbol{F}(\boldsymbol{x}), \forall \boldsymbol{x} \in \mathcal{S}. \tag{1.59}$$

Such functions are known as **conserved quantities, invariants,** or **first integrals.** The first two terms follow from the fact that the flow conserves the initial value of the function, which is hence invariant, whereas the third term is a consequence of their use in reducing the dimension of a dynamical system.

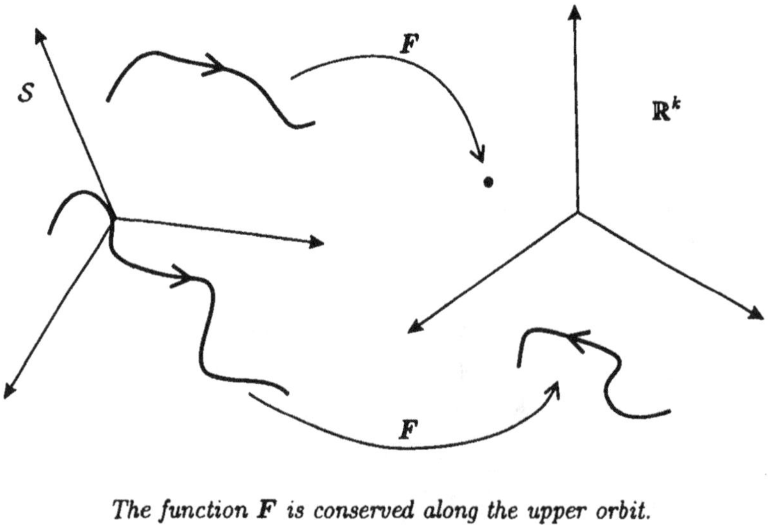

The function $\boldsymbol{F}$ *is conserved along the upper orbit.*

The presence of a conserved quantity, $F : S \to \mathbb{R}$, implies that state space can be partitioned into invariant submanifolds (level sets) on which the function F takes on a constant value. The continuity of the flow and the smoothness properties of F then imply that the stable and unstable manifolds of a stationary point must lie in the same submanifold as the stationary point itself. This follows from the fact that any initial state in the stable (unstable) manifold gets arbitrarily close to the stationary point in positive (negative) time. Similarly, the stable and unstable manifolds of hyperbolic periodic orbits, etc., also lie in the same submanifolds as these orbits.

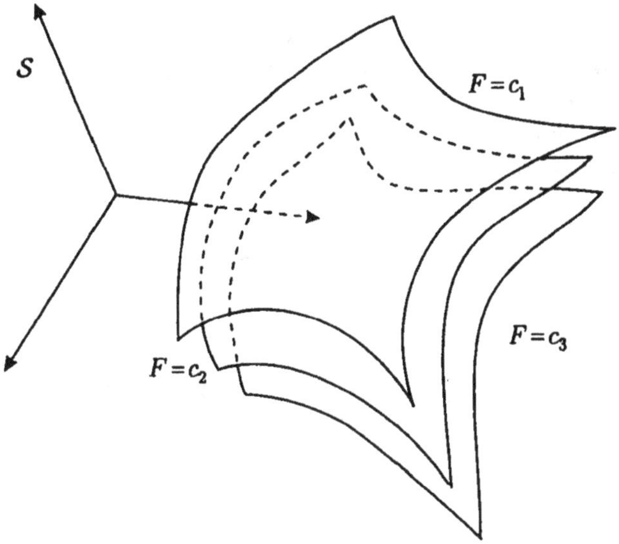

The conserved quantity F separates the flow into invariant submanifolds, F = constant.

If we are able to globally partition state space into invariant manifolds as discussed in the previous paragraph, then the analysis of the dynamical system is conveniently reduced to evolution on a submanifold. In this sense, the conserved quantities are known as isolating integrals. If such a partition is only possible locally, then the conserved quantities are not isolating and only limited information is available through them. For example, invariant submanifolds might exist on which the flow preserves the value of a function F, which, nevertheless, is not conserved for motion outside this submanifold. We will see examples of both cases in the coming chapters.

1.2 Hamiltonian systems

1.2.1 The equations of motion

The analysis of the behavior of general, continuous dynamical systems is made difficult by the complexity of the vector field and its high-dimensionality. It is, therefore, more than simply convenient if the evolution rule can be derived in terms of a lower-dimensional object, preferably a scalar function on state space. One such possibility is the **gradient** flow, which is defined through a **potential** function V by

$$\dot{x} = -\frac{\partial V}{\partial x}. \tag{1.60}$$

The negative sign in front of the differential is a matter of tradition and standard, although different sign conventions can sometimes be found. Potential, or gradient, flows are, for example, found in mathematical fluid mechanics. Their dynamics are simply deduced from the properties of the potential function and the analysis is consequently greatly simplified.

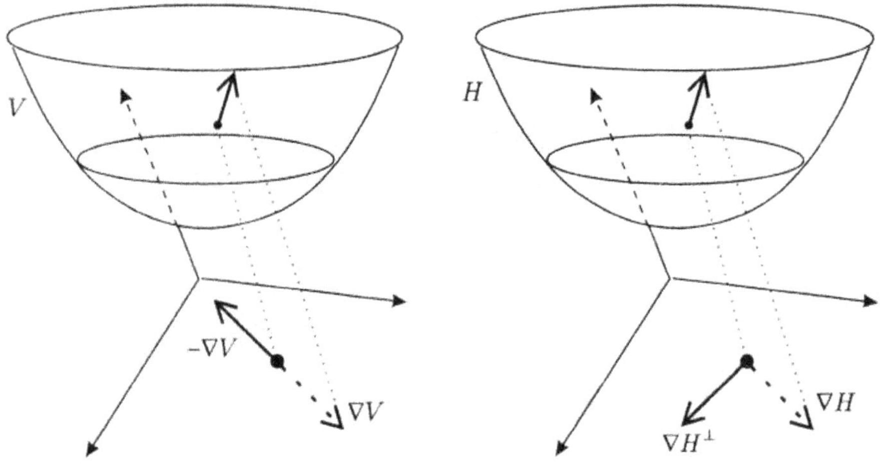

In the gradient flow (left panel) the vector field is parallel to the gradient of the potential function. In the Hamiltonian case (right panel), it is perpendicular to the gradient of the Hamiltonian.

Another, and for our purposes more relevant, case is that of a **Hamiltonian system**. Here, the flow is defined through the introduction of a **Hamiltonian (function)**, $H : \mathcal{S} \times \mathcal{R} \to \mathcal{R}$, such that

$$\dot{q} = \frac{\partial H}{\partial p}(q,p,t), \quad \dot{p} = -\frac{\partial H}{\partial q}(q,p,t) \tag{1.61}$$

where $\boldsymbol{x} = (\boldsymbol{q}, \boldsymbol{p}) \in S$ and $\boldsymbol{q}, \boldsymbol{p}$ are m-dimensional vectors, where m is the number of **degrees-of-freedom** of the Hamiltonian system. Evidently, Hamiltonian systems are even-dimensional. The dynamical system (1.61) is said to be in **canonical form** and the equations are often referred to as the **equations of motion**. The coordinates $\boldsymbol{q}$ and $\boldsymbol{p}$ are known as the **position** and **momentum coordinates** and when they satisfy equations of the form of Eq. (1.61), they are known as each others' **conjugates**.

For simplicity of notation, we shall find it useful to introduce a few auxiliary operators. The first is the symplectic identity matrix, $\boldsymbol{J}$, defined by

$$\boldsymbol{J} = \begin{pmatrix} \boldsymbol{0} & \boldsymbol{I} \\ -\boldsymbol{I} & \boldsymbol{0} \end{pmatrix}, \tag{1.62}$$

where $\boldsymbol{0}$ and $\boldsymbol{I}$ are the zero and unit matrix, respectively, of size $m \times m$. The canonical equations of motion can then be written as

$$\dot{\boldsymbol{x}} = \boldsymbol{J} \boldsymbol{D} H(\boldsymbol{x}, t). \tag{1.63}$$

Here, $\boldsymbol{D}$ denotes the differentiation operator and $\boldsymbol{D}H$ can be thought of as simply the vector of first partial derivatives.

Yet another way of writing the equations of motion is provided by the Poisson bracket operation

$$\{\cdot, \cdot\} : \mathcal{F} \times \mathcal{F} \to \mathcal{F} \tag{1.64}$$

on the function space $\mathcal{F}$ defined by

$$\{F, G\} = \frac{\partial F}{\partial \boldsymbol{q}} \cdot \frac{\partial G}{\partial \boldsymbol{p}} - \frac{\partial F}{\partial \boldsymbol{p}} \cdot \frac{\partial G}{\partial \boldsymbol{q}}. \tag{1.65}$$

The definition implies that

$$\{\boldsymbol{q}, H\} = \frac{\partial H}{\partial \boldsymbol{p}} = \dot{\boldsymbol{q}}, \quad \{\boldsymbol{p}, H\} = -\frac{\partial H}{\partial \boldsymbol{q}} = \dot{\boldsymbol{p}}. \tag{1.66}$$

For an arbitrary, real-valued function, $F(\boldsymbol{q}, \boldsymbol{p}, t)$, its evolution follows from

$$\dot{F} = \frac{\partial F}{\partial \boldsymbol{q}} \cdot \dot{\boldsymbol{q}} + \frac{\partial F}{\partial \boldsymbol{p}} \cdot \dot{\boldsymbol{p}} + \frac{\partial F}{\partial t} = \{F, H\} + \frac{\partial F}{\partial t}. \tag{1.67}$$

Clearly, Eq. (1.66), i.e., the equations of motion, are of this type.

A convenient feature of the Hamiltonian formulation is the existence of a first integral in the autonomous case. In fact, Eq. (1.67) immediately implies that $\dot{H} = 0$, since $\{H, H\} = 0$. Thus, the Hamiltonian itself is conserved throughout the motion. As in the discussion of the previous section, it is consequently possible to separate the state space into invariant submanifolds to which the motion is constrained. Since the value of the Hamiltonian in many physical problems can be related to the total energy of the system, the invariant manifolds are often known as **energy manifolds**. We note that the fact that states lie in the same energy manifold does not imply that they lie on the same orbit, nor that the orbit of one can get arbitrarily close to the other. In fact, the dimension of the energy manifolds is only one less than the full space, and thus very complicated dynamics can exist in each such manifold.

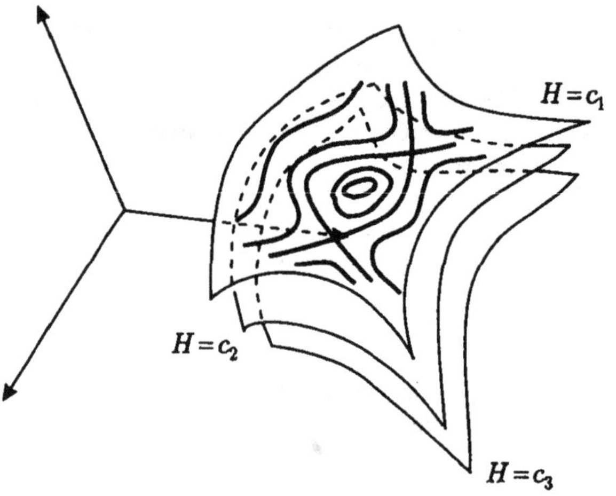

The Hamiltonian is a conserved quantity, hence the flow is constrained to invariant energy manifolds.

Illustration

Consider the two-degree-of-freedom Hamiltonian

$$H(\boldsymbol{q},\boldsymbol{p}) = \frac{1}{2}(p_1^2 + q_1^2) + \frac{1}{2}(p_2^2 + 2q_2^2) + q_1^2 q_2. \tag{1.68}$$

The energy manifolds are three-dimensional and flow on them is consequently difficult to visualize. A convenient compromise is achieved by slicing the energy manifold using the two-dimensional plane $q_2 = 0$*. The points of intersection of a given orbit on* $H =$ *constant with this plane serve as a projection of the full three-dimensional dynamics. In particular, we plot the* (q_1, p_1) *coordinates of such intersections when* $p_2 > 0$*. From the expression for* H*, it follows that these intersections lie inside a circle with radius equal to* $\sqrt{2H}$*.* •

In the previous section we argued for the possibility of transforming a non-autonomous dynamical system to an autonomous one by the introduction of auxiliary variables. We now outline how this is most appropriately done in the Hamiltonian case so as to allow the application of the results of the previous paragraph. Consider a time-dependent Hamiltonian $H(\boldsymbol{q},\boldsymbol{p},t)$. We introduce an extra position coordinate q_0 to take the place of t by letting it evolve according to

$$\dot{q}_0 = 1 \tag{1.69}$$

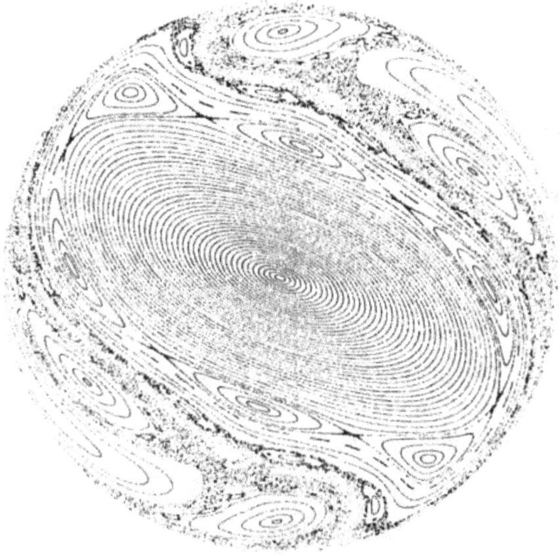

The intersections with the $q_2 = 0$ plane when $p_2 > 0$ and $H = 0.62$. Note the characteristic regions filled with closed curves and those with apparently stochastically distributed points.

with initial datum $q_0(t_0) = t_0$. For a canonical formulation we now require a conjugate momentum coordinate, p_0, so that the new **suspended** Hamiltonian $H_s(\boldsymbol{q}, q_0, \boldsymbol{p}, p_0)$ generates equations of motion for $\boldsymbol{q}, \boldsymbol{p}$ equivalent to those generated by H and for q_0 equal to Eq. (1.69), i.e.,

$$\frac{\partial H_s}{\partial \boldsymbol{q}} = \frac{\partial H}{\partial \boldsymbol{q}}, \quad \frac{\partial H_s}{\partial \boldsymbol{p}} = \frac{\partial H}{\partial \boldsymbol{p}}, \text{ and } \frac{\partial H_s}{\partial p_0} = 1. \tag{1.70}$$

The above conditions are trivially satisfied by the suspended Hamiltonian

$$H_s(\boldsymbol{q}, q_0, \boldsymbol{p}, p_0) = H(\boldsymbol{q}, \boldsymbol{p}, q_0) + p_0 \tag{1.71}$$

from which it immediately follows that

$$\dot{p}_0 = -\frac{\partial H}{\partial q_0} = -\frac{\partial H}{\partial t} = -\dot{H} \tag{1.72}$$

i.e., $H + p_0 = H_s$ is conserved. Thus, the introduction of the additional degree-of-freedom is accompanied by a new first integral. Since the initial condition on p_0 is arbitrary, we restrict attention to the zero-energy submanifold corresponding to the choice of initial condition

$$p_0(t_0) = -H(\boldsymbol{q}(t_0), \boldsymbol{p}(t_0), t_0). \tag{1.73}$$

It then follows that $p_0 \equiv -H(\boldsymbol{q}, \boldsymbol{p}, t)$.

Finally, we note that if the Hamiltonian is independent of a particular position coordinate, then the conjugate momentum is an invariant of the flow. If this happens, the position coordinate is said to be **cyclic**. In the autonomous case, the q_0 coordinate is clearly cyclic for the suspended Hamiltonian. For natural reasons, the presence of cyclic coordinates greatly reduces the complexity of the dynamics.

1.2.2 Stability of stationary points

The special form of the equations of motion limits the types of stationary points that can occur for Hamiltonian systems. Linearization of the equations of motion around a given stationary point at $\bar{\boldsymbol{x}}$ yields

$$\dot{\boldsymbol{y}} = \boldsymbol{J}\boldsymbol{D}^2 H(\bar{\boldsymbol{x}})\boldsymbol{y}, \tag{1.74}$$

where $\boldsymbol{D}^2 H$ is the Hessian of the Hamiltonian

$$\boldsymbol{D}^2 H = \begin{pmatrix} \frac{\partial^2 H}{\partial q^2} & \frac{\partial^2 H}{\partial q \partial p} \\ \frac{\partial^2 H}{\partial p \partial q} & \frac{\partial^2 H}{\partial p^2} \end{pmatrix}. \tag{1.75}$$

We now prove the following assertion: *If λ is an eigenvalue of $\boldsymbol{A} = \boldsymbol{J}\boldsymbol{D}^2 H$, then $-\lambda$ and $\pm\lambda^*$ are also eigenvalues.*

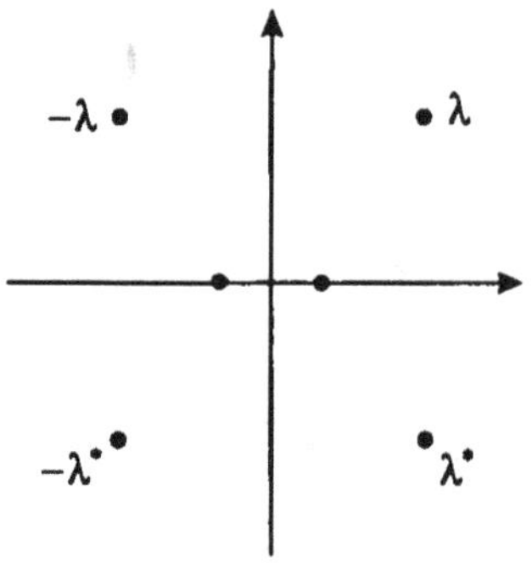

The eigenvalues corresponding to a stationary point of a Hamiltonian flow occur in sets, $\lambda, -\lambda, \lambda^*, -\lambda^*$.

In fact, since the matrix is real, all eigenvalues come in conjugate pairs, i.e., λ and λ^*. It remains to show that there exists a vector $\boldsymbol{w}$ such that

$$\boldsymbol{A}\boldsymbol{w} = -\lambda \boldsymbol{w}, \tag{1.76}$$

provided

$$Av = \lambda v \tag{1.77}$$

for some vector v. This follows from the symmetry of the second partial derivatives:

$$D^2 H = (D^2 H)^T, \tag{1.78}$$

where ^{T} denotes transpose. Then

$$JA + A^T J = 0, \tag{1.79}$$

since $-JJ = J^T J = I$. Thus, Eq. (1.77) implies

$$JAv = \lambda Jv \tag{1.80}$$

or

$$A^T Jv = -\lambda Jv, \tag{1.81}$$

i.e., A^T has the eigenvalue $-\lambda$. But A and A^T have the same eigenvalues and the claim thus follows.

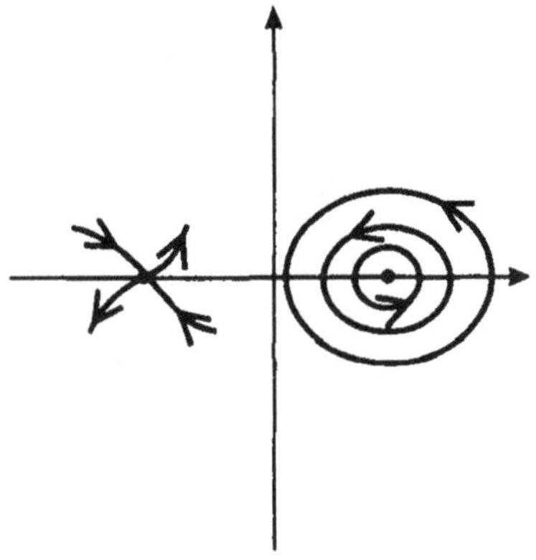

Stationary points of Hamiltonian flows are either saddle points or degenerate.

Thus, all hyperbolic equilibria of Hamiltonian systems are saddles; sinks and sources cannot occur. Linear stability can occur only for the case in which all eigenvalues are purely imaginary. However, as noticed previously, in this situation the Hartman-Grobman theorem fails to yield definite predictions for nonlinear stability. Hence, in the Hamiltonian case, the results of a linear analysis can be carried to the nonlinear system only in the case of unstable fixed points. These conclusions have fundamental implications on the type of dynamics observed in Hamiltonian systems, as will be discussed in later chapters. We turn now to the class of physical problems expressible within the Hamiltonian context known as classical mechanics.

1.3 Classical mechanics

Classical mechanics in its original form concerns itself with the dynamical description of particles and bodies under the influence of internal and external interactions. Several mathematical formulations for the necessary evolution equations are available, the original being the Newtonian equations of motion

$$\boldsymbol{a} = \frac{\boldsymbol{F}}{m} \tag{1.82}$$

for the acceleration, $\boldsymbol{a}$, of a particle of mass m under the influence of an external force, $\boldsymbol{F}$. For our purposes, the most convenient formulation is provided by the **modified Hamilton's principle,**

$$\delta \int_{t_1}^{t_2} (\boldsymbol{p} \cdot \dot{\boldsymbol{q}} - H(\boldsymbol{p}, \boldsymbol{q}, t))\, \mathrm{d}t = 0, \tag{1.83}$$

which should hold true for any choice of coordinates $\boldsymbol{p}, \boldsymbol{q}$. (The standard Hamilton's principle is the condition

$$\delta \int_{t_1}^{t_2} L(\boldsymbol{q}, \dot{\boldsymbol{q}}, t) \mathrm{d}t = 0$$

where L is the Lagrangian of the system.) The variation in the above expression is taken over all motions in state space for which

$$\delta p_i = \delta q_i = 0, \forall i \tag{1.84}$$

at $t = t_1, t_2$. The function H is suggestively known as the Hamiltonian for the

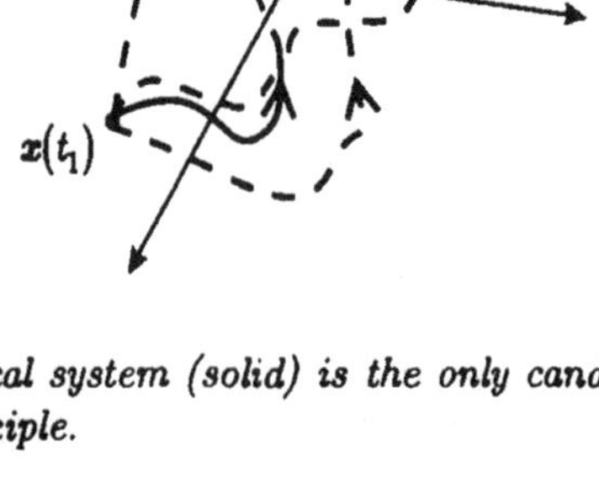

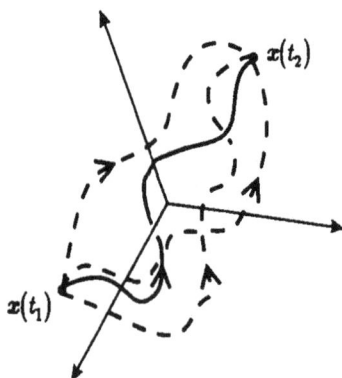

The flow of the mechanical system (solid) is the only candidate orbit to satisfy the modified Hamilton's principle.

mechanical system at hand. Since the time interval is kept fixed in the variation in Eq. (1.83), we obtain

$$0 = \int_{t_1}^{t_2} \left(\delta \boldsymbol{p} \cdot \dot{\boldsymbol{q}} + \boldsymbol{p} \cdot \delta \dot{\boldsymbol{q}} - \frac{\partial H}{\partial \boldsymbol{q}} \cdot \delta \boldsymbol{q} - \frac{\partial H}{\partial \boldsymbol{p}} \cdot \delta \boldsymbol{p} \right) \mathrm{d}t$$

$$= [\boldsymbol{p} \cdot \delta \boldsymbol{q}]_{t_1}^{t_2} + \int_{t_1}^{t_2} \left(\delta \boldsymbol{p} \cdot \left[\dot{\boldsymbol{q}} - \frac{\partial H}{\partial \boldsymbol{p}} \right] - \delta \boldsymbol{q} \cdot \left[\dot{\boldsymbol{p}} + \frac{\partial H}{\partial \boldsymbol{q}} \right] \right) \mathrm{d}t \tag{1.85}$$

where we have performed a partial integration of the term $\boldsymbol{p} \cdot \delta \dot{\boldsymbol{q}}$ using the fact

$$\delta \frac{d}{dt} \boldsymbol{q} = \frac{d}{dt} \delta \boldsymbol{q}. \tag{1.86}$$

From Eq. (1.84) it now follows that the first term on the right-hand side of Eq. (1.85) vanishes. Furthermore, since the variations $\delta \boldsymbol{p}$ and $\delta \boldsymbol{q}$ are independent, the modified Hamilton's principle implies

$$\dot{\boldsymbol{q}} = \frac{\partial H}{\partial \boldsymbol{p}}, \qquad \dot{\boldsymbol{p}} = -\frac{\partial H}{\partial \boldsymbol{q}}, \tag{1.87}$$

which we recognize as identical to Eq. (1.61). Here $\boldsymbol{q}$ is known as the **generalized position vector** and $\boldsymbol{p}$ as the **generalized momentum vector**. By the terminology of the previous section, we talk of conjugate position and momentum coordinates.

From the discussion above, it is now clear that the description of a mechanical system as a dynamical system is given through the specification of the Hamiltonian, H. For conservative systems, it is normally given by

$$H = T + V, \tag{1.88}$$

where T is the kinetic energy and V a time-independent potential energy. In Cartesian coordinates we then write

$$T = \frac{1}{2m} \boldsymbol{p} \cdot \boldsymbol{p}, \qquad V = V(\boldsymbol{q}) \tag{1.89}$$

from which Hamilton's equation give

$$\dot{\boldsymbol{q}} = \frac{1}{m} \boldsymbol{p}, \qquad \dot{\boldsymbol{p}} = -\frac{\partial V}{\partial \boldsymbol{q}}. \tag{1.90}$$

Thus, we find

$$\boldsymbol{p} = m \dot{\boldsymbol{q}} \tag{1.91}$$

and hence

$$m \ddot{\boldsymbol{q}} = -\frac{\partial V}{\partial \boldsymbol{q}}, \tag{1.92}$$

which are Newton's equations for a particle subject only to potential forces. Since $T + V = E$, where E is the total energy of the system, we see that $H \equiv E$, and the constancy of the energy follows from the fact that H is conserved.

In the following section, we consider the specific class of mechanical systems given by the motion of astronomical objects under the influence of gravitational and other forces.

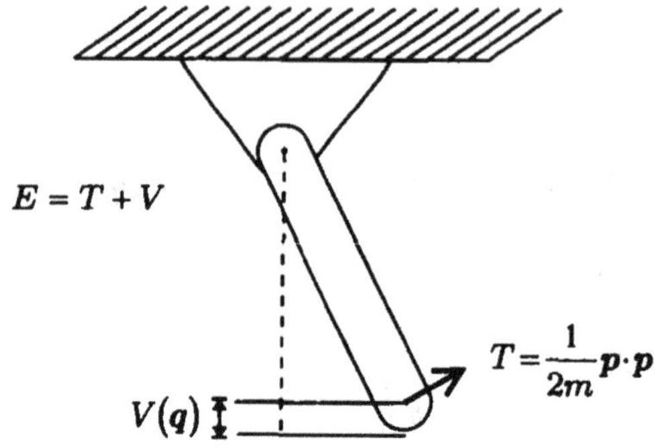

In the mechanical system of a simple pendulum, the energy is conserved, but undulates between its kinetic and potential components.

1.4 Celestial mechanics

The phenomenological laws of Kepler described in the introduction were given a solid theoretical background with the **Newtonian law of gravity**

$$\boldsymbol{F}_{grav} = G\frac{m_1 m_2}{r^3}(\boldsymbol{r}_1 - \boldsymbol{r}_2) \tag{1.93}$$

describing the gravitational force experienced by a particle of mass m_2 at $\boldsymbol{r}_2$ interacting with the particle of mass m_1 at $\boldsymbol{r}_1$. Here $r = |\boldsymbol{r}_1 - \boldsymbol{r}_2|$ and G is the gravitational

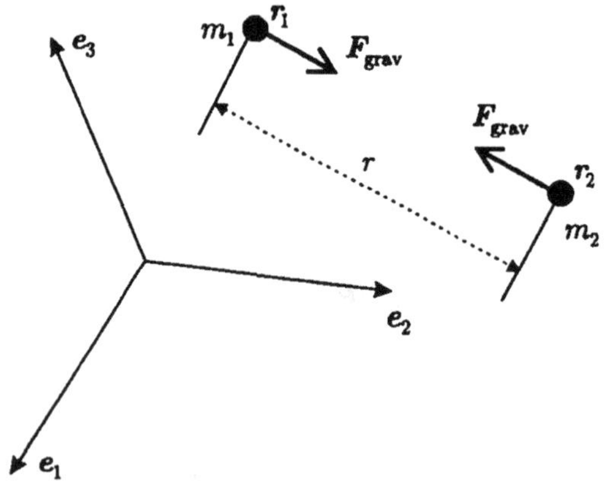

The gravitational interaction between two particles depends on their masses and relative distance alone.

constant $\approx 6.67 \cdot 10^{-11}$ Nm^{2}kg^{-2} in SI units. Although Eq. (1.93) is defined for particle-particle interactions, it is, in fact, possible to show that it also holds when two spherically symmetric bodies of finite extent are studied.

Consider the gravitational interaction between a particle of mass m_2 at $(0,0,z)$ and a spherical body of radius R, constant density ρ, and center at the origin. From symmetry it is clear that $\boldsymbol{F}_{grav} = F_1\boldsymbol{e}_3$. With the slicing suggested below, Eq. (1.93) yields

$$F_1 = -Gm_2\rho \int_{-R}^{R}\int_0^{2\pi}\int_0^{\sqrt{R^2-\zeta^2}} \frac{z-\zeta}{\left[(z-\zeta)^2+\xi^2\right]^{3/2}}\xi d\xi d\theta d\zeta = -G\frac{m_1 m_2}{z^2}, \quad (1.94)$$

where $m_1 = \frac{4\pi R^3}{3}\rho$ is the mass of the spherical body. Thus, each point in a spherical body experiences the gravitational attraction from a second spherical body as if the mass of the latter were concentrated at its center. Finally, integrating over all particles constituting the first body again yields Eq. (1.93). While celestial bodies in general have nonspherical shapes due to their rotation, the spherical symmetry assumption is an adequate first approximation and quite sufficient when the distance between the bodies is large. Changes in the gravitational interactions due to asphericity will not be considered in this text.

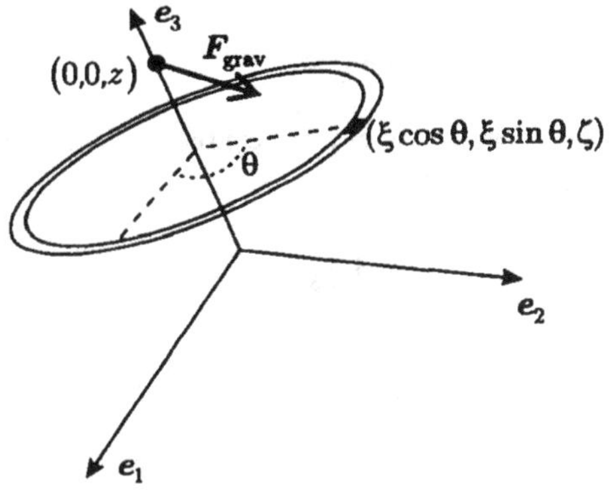

Appropriate slicing for evaluating the gravitational attraction between two spherical bodies.

We now consider the mechanical system given by two gravitationally interacting bodies, the second of whom is further subject to an external acceleration, $\boldsymbol{P}$. Newton's equation of motion then gives

$$\ddot{\boldsymbol{r}}_1 = G\frac{m_2}{r^3}(\boldsymbol{r}_2 - \boldsymbol{r}_1) \quad (1.95)$$

$$\ddot{\boldsymbol{r}}_2 = G\frac{m_1}{r^3}(\boldsymbol{r}_1 - \boldsymbol{r}_2) + \boldsymbol{P}. \quad (1.96)$$

In the general two-body problem, it is often desirable to describe the motion of the second body, the satellite, in a reference frame in which the first body, known as the primary, is fixed. In the solar system this is motivated by the fact that the relative masses of all the planets are much smaller than the sun, and the latter hence experiences much smaller accelerations.

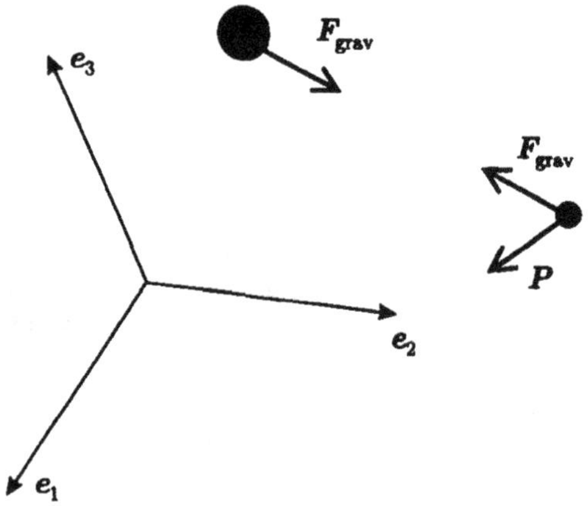

A perturbed two-body problem.

We transform to a coordinate system centered on the primary body by introducing the new coordinates $\boldsymbol{r} = \boldsymbol{r}_2 - \boldsymbol{r}_1$. From Eqs. (1.95-1.96) we then obtain

$$\ddot{\boldsymbol{r}} + \frac{\mu}{r^3}\boldsymbol{r} = \boldsymbol{P} \tag{1.97}$$

where $\mu = G(m_1 + m_2)$. When the perturbing acceleration, $\boldsymbol{P}$, is derived from a (possibly time-dependent) potential

$$\boldsymbol{P} = -\frac{\partial V}{\partial \boldsymbol{r}}, \tag{1.98}$$

then Eq. (1.97) is equivalent to the Hamiltonian

$$H(\boldsymbol{p}, \boldsymbol{r}, t) = \frac{1}{2}\boldsymbol{p} \cdot \boldsymbol{p} - \frac{\mu}{r} + V(\boldsymbol{r}, t). \tag{1.99}$$

In the case of pure two-body motion, where $\boldsymbol{P} \equiv 0$, the equations of motion are analytically solvable. This is most easily shown in polar coordinates (ϱ, θ, z), which are related to the Cartesian coordinates (x, y, z) through

$$x = \varrho\cos\theta, \quad y = \varrho\sin\theta, \quad z = z. \tag{1.100}$$

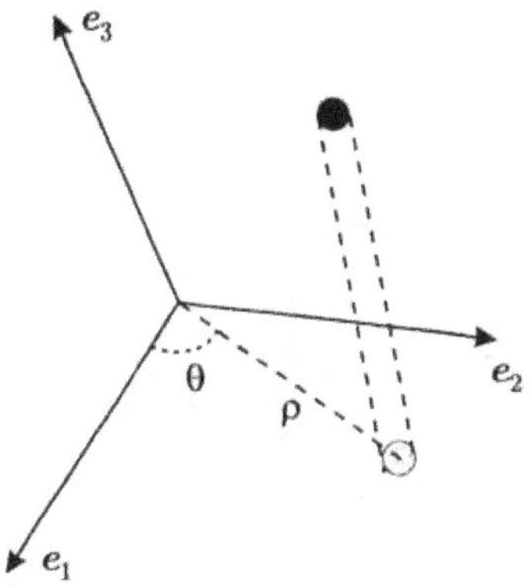

Polar coordinates in the reference frame centered on the primary.

To retain the canonical form of the equations of motion, we introduce the conjugate transformation for the momenta (we show in the next chapter how this is found),

$$p_x = p_\varrho \cos\theta - \frac{1}{\varrho} p_\theta \sin\theta, \quad p_y = p_\varrho \sin\theta + \frac{1}{\varrho} p_\theta \cos\theta, \quad p_z = p_z \tag{1.101}$$

and thus

$$H = \frac{1}{2}\left(p_\varrho^2 + \frac{p_\theta^2}{\varrho^2} + p_z^2\right) - \frac{\mu}{r}, \tag{1.102}$$

where $r = \sqrt{\varrho^2 + z^2}$. The z-component of Hamilton's equations then yields

$$\ddot{z} = -\frac{\mu}{r^3} z. \tag{1.103}$$

Hence, if the coordinate system is chosen so that the particle's motion initially lies in the $z = 0$ plane, then it will remain there for all time and hence $p_z \equiv 0$. Consequently, the gravitational potential is only dependent on ϱ. Furthermore, since θ is cyclic, we find

$$p_\theta = l, \tag{1.104}$$

where l is called the angular momentum of the particle.

The ϱ and θ components of Hamilton's equations

$$\dot{\varrho} = p_\varrho = \sqrt{2E - \frac{l^2}{\varrho^2} + 2\frac{\mu}{\varrho}}, \tag{1.105}$$

where $H = E =$ constant, since H is conservative, and

$$\dot{\theta} = \frac{l}{\varrho^2} \tag{1.106}$$

can be combined to give

$$\frac{d\varrho}{d\theta} = \frac{\dot{\varrho}}{\dot{\theta}} = \frac{\varrho^2\sqrt{2E - \frac{l^2}{\varrho^2} + 2\frac{\mu}{\varrho}}}{l}. \tag{1.107}$$

Solving this differential equation for $\varrho(\theta)$ one finds

$$\varrho = \frac{a(1-e^2)}{1+e\cos\theta} \tag{1.108}$$

where

$$l = \sqrt{\mu a(1-e^2)}, \qquad E = -\frac{\mu}{2a}. \tag{1.109}$$

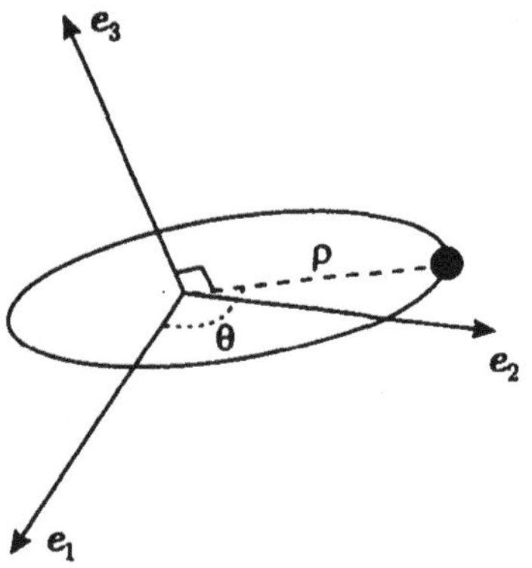

The only bounded solutions are ellipses with one focus on the primary.

For $e \le 1$ Eq. (1.108) is the equation for an ellipse with semimajor axis a and eccentricity e, where the origin is located at one of the foci. For all other values of e the motion is unbounded. Thus, in the reference frame of an observer moving with the sun, Newton's law of gravity implies that the bounded motion of the planets around the sun satisfies the first law of Kepler. From Eqs. (1.95-1.96) we see that $\boldsymbol{P} = \boldsymbol{0}$ implies that $m_1\ddot{\boldsymbol{r}}_1 + m_2\ddot{\boldsymbol{r}}_2 = \boldsymbol{0}$, i.e., the center of mass moves with constant velocity. Moreover,

$$\boldsymbol{r}_1 - \frac{m_1\boldsymbol{r}_1 + m_2\boldsymbol{r}_2}{m_1+m_2} = -\frac{m_2}{m_1+m_2}\boldsymbol{r} \tag{1.110}$$

and

$$\boldsymbol{r}_2 - \frac{m_1\boldsymbol{r}_1 + m_2\boldsymbol{r}_2}{m_1+m_2} = \frac{m_1}{m_1+m_2}\boldsymbol{r}. \tag{1.111}$$

Hence, in the original reference frame, the primary and satellite move on elliptical orbits with their common center of mass at one of the foci. If we recall that the sun's mass outstrips that of any other solar system object by far, and hence that the common center of mass of the sun and any other object lies somewhere inside the sun, this confirms Kepler's first law.

That the relative motion is restricted to a plane has already been shown. The remaining claim of the second of Kepler's laws about the constancy of the rate of change of the area $A(t)$ swept by the position vector follows from Eq. (1.106), since

$$\dot{A} = \frac{\varrho^2\dot{\theta}}{2} = \frac{l}{2}. \tag{1.112}$$

Finally, substituting Eq. (1.108) into Eq. (1.106) yields

$$dt = \frac{[a(1-e^2)]^{\frac{3}{2}}}{\sqrt{\mu}(1+e\cos\theta)^2}d\theta. \tag{1.113}$$

Integrating both sides of this equality over one period $t \in [0, T]$, $\theta \in [0, 2\pi]$ then yields

$$T = \frac{2\pi a^{\frac{3}{2}}}{\sqrt{\mu}}. \tag{1.114}$$

In the solar system, the sun's mass is so overwhelmingly large that μ is essentially independent of the planet under consideration. Consequently, $T^2 \sim a^3$, from which Kepler's third law follows.

In later chapters in this book we shall be concerned with a particular perturbation on the two-body motion described above, resulting from the gravitational influence of a third body and the radiation pressure originating in the absorption and re-emission of solar photons by the satellite. For small perturbations it is a fair assumption that the motion will in some sense be close to the unperturbed one. Whether this is indeed the case will be studied further in the next chapter. The validity of the assumption for sufficiently short intervals of time forms the basis for the introduction of the osculating elements in celestial mechanics. (Naturally, short is a very relative measure. The dynamical astronomer will often accept hundreds of thousands of years as a reasonable short time.) These are related to the geometric properties of an auxiliary ellipse, the osculating orbit; and their use allows for easily interpreted geometric statements to be made about the perturbed orbits in terms of familiar concepts. While the osculating elements have found much application in a large body of celestial mechanics work, we will refrain from their introduction in this text. The choice of variables in which to describe the effects of the perturbation will instead be dictated by the problem at hand.

But we are running ahead of ourselves. Many properties of Hamiltonian systems in particular, and dynamical systems in general, need to be illuminated before specific examples can be considered. It is to this task that we now turn.

1.5 Notes and references

An often-journeyed route is likely to be depicted in many a traveler's guide. So too with the areas of mathematics and physics discussed in this chapter. Thus, no attempt at completeness has been made with regard to the references below.

Dynamical systems

With respect to dynamical systems as a field, this chapter has attempted to introduce the novice to an established terminology. There is an abundance of introductory texts to dynamical systems of which we particularly mention those by Arrowsmith & Place[1], Guckenheimer & Holmes[2], and Hirsch & Smale[3]. With varying emphasis, these references are complementary and provide a comprehensive presentation of dynamical systems theory.

1. D. K. Arrowsmith and C. M. Place, *An Introduction to Dynamical Systems* (Cambridge University Press, 1990).
2. J. Guckenheimer and P. Holmes, *Nonlinear Oscillations, Dynamical Systems, and Bifurcations of Vector Fields, 2nd edition* (Springer Verlag, 1990).
3. M. W. Hirsch and S. Smale, *Differential Equations, Dynamical Systems, and Linear Algebra* (Academic Press, 1974).

Hamiltonian systems and classical mechanics

The detail and complexity of Hamiltonian systems has only been briefly touched upon in this chapter. Introductory texts with similar material are perhaps as abundant as those on differential equations in general. The classic texts of Goldberg[1] and Whittaker[2] provide in-depth presentations of the subject. An interesting point of view is further given in Tolman[3]. The analytical methods of classical mechanics are similarly a subject of a large literature, much of which overlaps that of Hamiltonian systems. In addition to the references above, we mention that of Arnold[4].

1. H. Goldstein, *Classical Mechanics* (Addison-Wesley Publishing Co., 1959).
2. E. T. Whittaker, *A Treatise on the Analytical Dynamics of Particles and Rigid Bodies: with an Introduction to the Problem of Three Bodies., 4th edition* (Cambridge University Press, 1988).
3. R. C. Tolman, *The Principles of Statistical Mechanics* (Dover, 1979).
4. V. I. Arnold, *Mathematical Methods of Classical Mechanics* (Springer Verlag, 1989).

Celestial mechanics

The fascinating subject of celestial mechanics has grown with the development of the tools of dynamical systems. Similarly, many of the texts above contain more or less in-depth discussions of celestial mechanics applications. The classical approach to the subject is thoroughly described in the books of Brouwer & Clemence[1] and Danby[2].

1. D. Brouwer and G. M. Clemence, *Methods of Celestial Mechanics* (Academic Press, 1961).
2. J. M. A. Danby, *Fundamentals of Celestial Mechanics, 2nd edition* (Willmann-Bell, 1988).

Chapter 2
HAMILTONIAN SYSTEMS

The possibility to extrapolate from the near to the far, from the small to the large, from the simple to the complex, is what makes most scientific effort a worthwhile endeavor. Complicated physical systems are stripped of all but the most evident causes and the analysis of their effects is then assumed to bear relevance on the original system. At the heart lies the premise that, given that any additional causes are sufficiently small, then, indeed, the actual system behavior will be close to the simplified one. We speak of **structurally stable** behavior when such conclusions can be drawn. It is the task of the investigator to ascertain whether a particular behavior truly is structurally stable or if, instead, it is highly sensitive to additional perturbations. In the former case, much insight will have been gained into the original problem. In the latter, a return to the proverbial drawing board is inevitable.

A highly structurally unstable situation is obtained by placing a ball on a rough plane perpendicular to a gravitational field. Clearly, a small perturbation of the angle of the table will cause the ball to roll off.

In the previous chapter it was suggested that classical mechanics, and particularly celestial mechanics, could be conveniently expressed using Hamiltonian dynamics. We showed how, with the aid of conserved quantities in the two-body problem, we were able to effectuate its complete solution and a description of its global dynamics. We shall generalize these ideas in this chapter and apply them to an arbitrary Hamiltonian system. Of fundamental importance will be the distinction between systems whose dynamics are, in some sense, regular and predictable, and those whose behavior is largely stochastic. Section 1 is devoted to describing some of the ideas underlying the notion of regular motion.

The Hamiltonian formulation is a very powerful one. Retaining the symmetries afforded by it while attempting to simplify the description of a given dynamical system is an ever-present goal. This can indeed be achieved through the use of a special class of transformations, the theory of which is described in detail in Section 2. Armed with these transformations, Section 3 shows how to reduce the issue from whether a given system is of the regular or irregular type, to showing the existence of a solution to a particular partial differential equation. When a solution exists, the system is known as integrable and all its behavior is of the regular, predictable type.

Having simplified and modeled a given physical system with an integrable Hamiltonian system, it is thus natural to ask whether the resulting behavior is structurally stable. A more restricted question is whether structural stability can even be guaranteed within the class of Hamiltonian systems. In other words, do two Hamiltonian systems, one of which is integrable, exhibit the same qualitative behavior provided they are close in some sense? The resounding answer provided by KAM theory and which we describe in Section 4 is "maybe." While some dynamics of the integrable problem are generally retained by nearby systems, others are not. Instead, regions of highly irregular and stochastic motion appear for which standard methods of analysis fail. The origin of these stochastic regions lies in the fate of stable and unstable manifolds of unstable periodic orbits in the flow. We will see similar effects, albeit not necessarily in this context, in the coming chapters.

2.1 Invariants

In the previous chapter we discussed the dynamical implications of invariants on the geometry of a flow. We have already seen that in an autonomous Hamiltonian system, one such invariant is provided by the Hamiltonian itself, since

$$\dot{H} = \{H, H\} = 0 \tag{2.1}$$

from Eq. (1.67). Consequently, the flow is restricted to submanifolds of state space given by $H = c$ for c constant. From Eq. (1.67) we further see that any other real-valued invariant of the flow, F, has to commute with H, i.e., $\{H, F\} = 0$. Such an invariant provides another separation of state space into invariant submanifolds. The intersection of one such submanifold with an energy manifold is also invariant. Only when the intersection is of lower dimension, i.e., when F is functionally independent of H, does F's existence allow the flow to be further simplified. We shall shortly pursue the question of reducing a given Hamiltonian flow using a sufficiently large set of independent invariants. But first we will introduce some convenient notation and important notions.

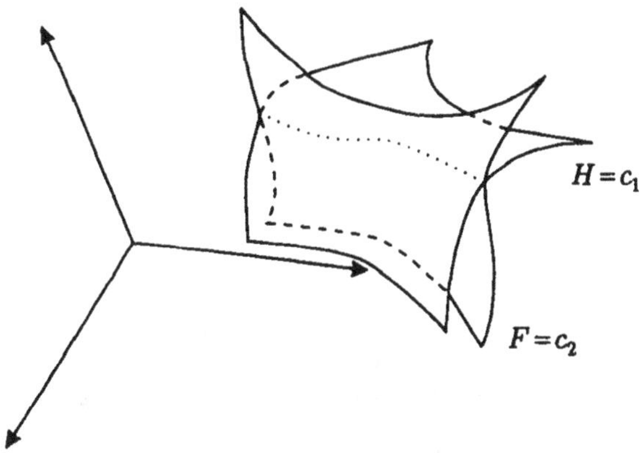

The intersections of submanifolds corresponding to constant value of invariants are also invariant.

2.1.1 *Poisson brackets*

We have previously defined the Poisson bracket of two functions, F and G, defined on state space as the function

$$\{F,G\} = \frac{\partial F}{\partial \boldsymbol{q}} \cdot \frac{\partial G}{\partial \boldsymbol{p}} - \frac{\partial F}{\partial \boldsymbol{p}} \cdot \frac{\partial G}{\partial \boldsymbol{q}}. \tag{2.2}$$

The Poisson bracket provided a compact notation for the time derivative of a function F of $\boldsymbol{q}$ and $\boldsymbol{p}$ under the flow generated by the Hamiltonian G. If the Poisson bracket of F and G vanishes identically, then F and G are said to Poisson commute. A convenient notation is provided by the Lie derivative operator $L_{\boldsymbol{X}}$ defined by

$$(L_{\boldsymbol{X}} F)(\boldsymbol{z}) = \left.\frac{\mathrm{d}}{\mathrm{d}t}\right|_{t=0} F(\boldsymbol{\phi}_{\boldsymbol{X}}(t, \boldsymbol{z})), \tag{2.3}$$

where F is a real-valued function, $\boldsymbol{X}$ is the vector field, and $\boldsymbol{\phi}_{\boldsymbol{X}}$ the corresponding flow. For a Hamiltonian vector field $\boldsymbol{X}_G$ corresponding to the Hamiltonian G the above becomes

$$(L_{\boldsymbol{X}_G} F)(\boldsymbol{q},\boldsymbol{p}) = \{F,G\}(\boldsymbol{q},\boldsymbol{p}). \tag{2.4}$$

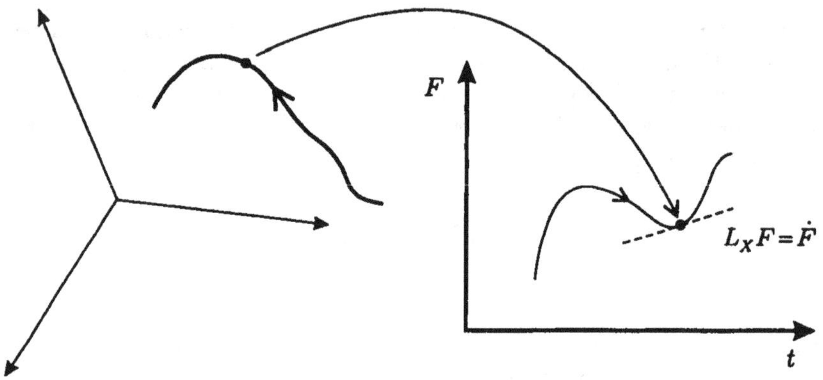

The Lie derivative of the function F at a point on an orbit is a measure of the rate of change of F along the orbit.

By expanding in partial derivatives the *Jacobi identity*

$$\{\{F,G\},H\} + \{\{G,H\},F\} + \{\{H,F\},G\} = 0 \tag{2.5}$$

is easily confirmed. In terms of the Lie operator this becomes

$$L_{X_{\{G,H\}}} F = \left(L_{X_H} L_{X_G} - L_{X_G} L_{X_H}\right) F. \tag{2.6}$$

If we define the commutator between two vector fields, $[\boldsymbol{X}, \boldsymbol{Y}]$, from

$$L_{[X,Y]} = L_X L_Y - L_Y L_X \tag{2.7}$$

we can rewrite Eq. (2.6) as

$$L_{X_{\{G,H\}}} = L_{[X_G, X_H]}. \tag{2.8}$$

In other words, the vector field corresponding to the Hamiltonian $\{G,H\}$ is given by the commutator $[\boldsymbol{X}_G, \boldsymbol{X}_H]$. Consequently, if G and H Poisson commute, then the commutator $[\boldsymbol{X}_G, \boldsymbol{X}_H]$ must vanish everywhere.

2.1.2 The commutator

In coordinate form the Lie operator becomes

$$L_X F = X_i F_{,i}. \tag{2.9}$$

Thus,

$$\left(L_X L_Y - L_Y L_X\right) F = \left(X_i Y_{j,i} - Y_i X_{j,i}\right) F_{,j} \tag{2.10}$$

or

$$[\boldsymbol{X}, \boldsymbol{Y}]_j = X_i Y_{j,i} - Y_i X_{j,i}. \tag{2.11}$$

Now consider the situation depicted below. Provided the vector field $\boldsymbol{X}$ is nonsingular at the point $\boldsymbol{A}$, there will be only one flow curve through each point in a neighborhood of $\boldsymbol{A}$. If $\boldsymbol{X}$ and $\boldsymbol{Y}$ are linearly independent, then similar restrictions on $\boldsymbol{Y}$ allow us to follow $\phi_{\boldsymbol{X}}(t_1, \boldsymbol{A})$ for time t_1 to the point $\boldsymbol{B}_1$, then $\phi_{\boldsymbol{Y}}(t_2, \boldsymbol{B}_1)$ for time t_2 to the point $\boldsymbol{C}_1$. Similarly, we can follow $\phi_{\boldsymbol{Y}}(t_2, \boldsymbol{A})$ to the point $\boldsymbol{B}_2$ and finally $\phi_{\boldsymbol{X}}(t_1, \boldsymbol{B}_2)$ to $\boldsymbol{C}_2$. In general, $\boldsymbol{C}_1$ and $\boldsymbol{C}_2$ do not coincide. For small t_1 and t_2, their separation can be shown to be approximately given by the commutator of $\boldsymbol{X}$ and $\boldsymbol{Y}$ at $\boldsymbol{A}$.

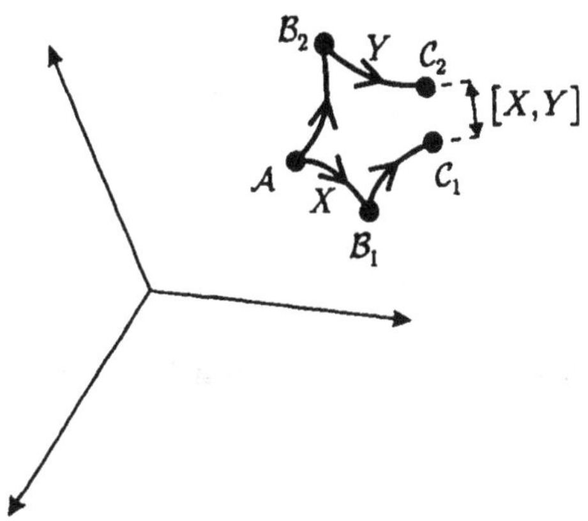

The commutator is a measure of the separation between states obtained by permuting the order of the flows.

In particular, we study the difference between the function F evaluated at the points $\boldsymbol{C}_1$ and $\boldsymbol{C}_2$. When F equals a state variable, this difference is simply the variable separation. We now find

$$\begin{aligned} F(\boldsymbol{C}_2) - F(\boldsymbol{C}_1) &= [F(\boldsymbol{C}_2) - F(\boldsymbol{B}_2)] + [F(\boldsymbol{B}_2) - F(\boldsymbol{A})] \\ &\quad + [F(\boldsymbol{A}) - F(\boldsymbol{B}_1)] + [F(\boldsymbol{B}_1) - F(\boldsymbol{C}_1)] \\ &= (L_{t_1\boldsymbol{X}}F)(\boldsymbol{B}_2) + (L_{t_2\boldsymbol{Y}}F)(\boldsymbol{A}) - (L_{t_1\boldsymbol{X}}F)(\boldsymbol{A}) - (L_{t_2\boldsymbol{Y}}F)(\boldsymbol{B}_1) + \text{h.o.t.} \\ &= (L_{\boldsymbol{Y}}L_{\boldsymbol{X}}F - L_{\boldsymbol{X}}L_{\boldsymbol{Y}}F)(\boldsymbol{A})t_1t_2 + \text{h.o.t.} \end{aligned} \tag{2.12}$$

where we have used $L_{a\boldsymbol{X}} = aL_{\boldsymbol{X}}$ and the higher order terms (h.o.t.) go to zero faster than the first term. Thus, we finally obtain

$$\frac{1}{t_1t_2}[F(\boldsymbol{C}_2) - F(\boldsymbol{C}_1)] = (L_{[\boldsymbol{X},\boldsymbol{Y}]}F)(\boldsymbol{A}) + o(|t_1, t_2|) \tag{2.13}$$

and so as $t_1, t_2 \to 0$ the ratio approaches $L_{[\boldsymbol{X},\boldsymbol{Y}]}F$. Hence, the vanishing of the commutator implies that the ratio

$$\frac{1}{t_1t_2}[\phi_{\boldsymbol{X}}(t_1, \phi_{\boldsymbol{Y}}(t_2, \boldsymbol{A})) - \phi_{\boldsymbol{Y}}(t_2, \phi_{\boldsymbol{X}}(t_1, \boldsymbol{A}))] \to 0 \text{ as } t_1, t_2 \to 0. \tag{2.14}$$

In fact, it is possible to show that if $[\boldsymbol{X}, \boldsymbol{Y}] \equiv 0$, then (2.14) vanishes for sufficiently small but nonzero t_1, t_2. If this happens, the flows ϕ_X and ϕ_Y are said to commute.

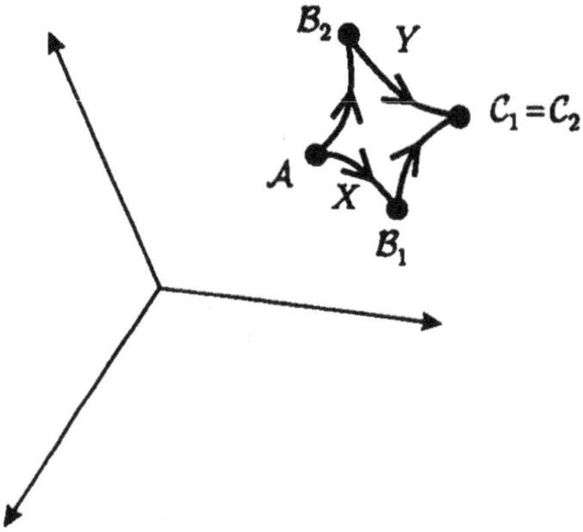

When the commutator vanishes, the order in which the flows are applied is unimportant, at least for small excursions.

2.1.3 Separability

We now assume that we have found an n-dimensional vector of independent invariants $\boldsymbol{F} = (F_1 = H, F_2, \ldots, F_n)$ that Poisson commute pairwise. Such a set of invariants is said to be **complete**. Since the invariants are independent, i.e., their gradients are linearly independent, the intersections of their level sets are n-dimensional invariant submanifolds, $S/\boldsymbol{F}$, of state space under the flow generated by the Hamiltonian H. We now consider the Hamiltonian vector field $\boldsymbol{X}_{F_i}$ generated by F_i. We then have

$$L_{\boldsymbol{X}_{F_i}} F_j = \{F_j, F_i\} = 0 \tag{2.15}$$

i.e., the invariants are also conserved under the flow of $\boldsymbol{X}_{F_i}$. In other words, the manifolds $S/\boldsymbol{F}$ are also invariant under the flow $\boldsymbol{X}_{F_i}$. Thus, the vector fields $\boldsymbol{X}_{F_i}$, $i = 1, \ldots, n$ are tangent to $S/\boldsymbol{F}$. Moreover, they are linearly independent and hence span the tangent space $S/\boldsymbol{F}$ at each point. Finally, the previous section shows that the flows $\phi_{\boldsymbol{X}_{F_i}}$, $i = 1, \ldots, n$ commute pairwise.

We now introduce a parameterization on $S/\boldsymbol{F}$ using the flows $\phi_{\boldsymbol{X}_{F_i}}$. In particular, consider the map

$$\phi_{\boldsymbol{X}_{F_1}}(t_1, \phi_{\boldsymbol{X}_{F_2}}(t_2, \ldots, z)) \ldots) \tag{2.16}$$

taking the point z to its image after flowing along $\boldsymbol{X}_{F_1}$ for time t_1, $\boldsymbol{X}_{F_2}$ for time t_2 etc. Since the flows commute, the subscripts can be arbitrarily reordered. If we fix a

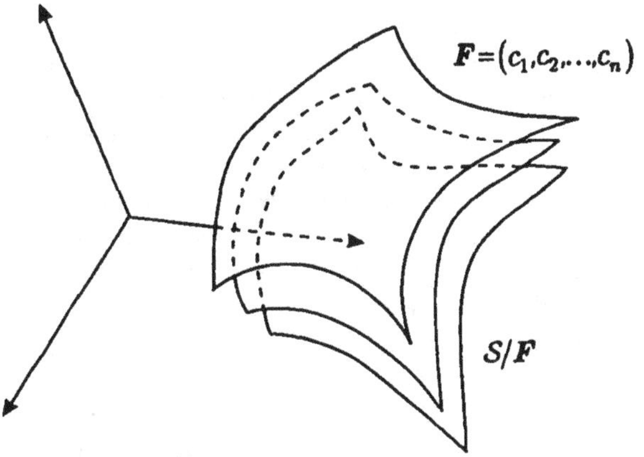

The collection of intersections of level sets for the invariants constitutes $S/\boldsymbol{F}$.

reference point $\mathcal{A}$, we obtain the map

$$\mathbb{R}^n \ni \boldsymbol{t} \mapsto \phi(\boldsymbol{t}, \mathcal{A}) \in S/\boldsymbol{F}, \tag{2.17}$$

where $\boldsymbol{t} = (t_1, t_2, \ldots, t_n)$ and ϕ is shorthand for the composition in Eq. (2.16). Note that $\phi(\boldsymbol{0}, \mathcal{A}) = \mathcal{A}$. Furthermore, since the flows commute, it follows that

$$\phi(\boldsymbol{t}, \phi(\boldsymbol{s}, \mathcal{A})) = \phi(\boldsymbol{t} + \boldsymbol{s}, \mathcal{A}), \forall \boldsymbol{t}, \boldsymbol{s} \tag{2.18}$$

i.e., $\phi(\cdot, \mathcal{A})$ is a group.

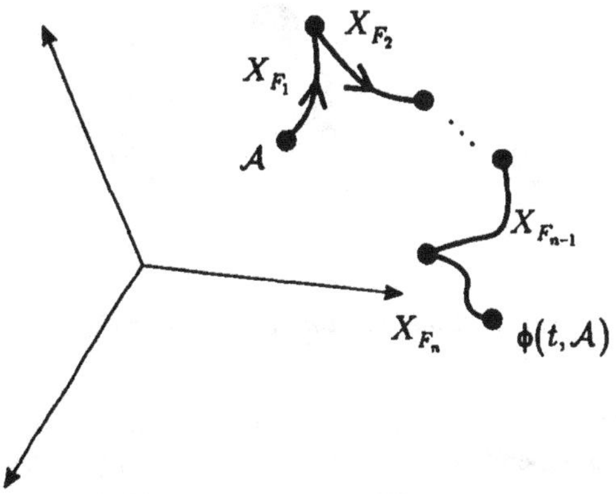

The map $\phi(\ \boldsymbol{t}, \mathcal{A})$ composes the flows corresponding to the independent invariants.

The commutativity of the flows implies that the matrix of partial derivatives of the components of ϕ, with respect to those of t at $\mathcal{A}$, has $X_{F_i}(\mathcal{A})$ as columns. Since the vector fields are linearly independent, it follows that the matrix has maximum rank and thus, by the implicit function theorem, that ϕ is one-to-one and onto with a differentiable inverse on a neighborhood of $\mathbf{0}$. In fact, ϕ is globally onto, since given a point $\mathcal{B} \neq \mathcal{A}$, we can connect the two points by a bounded curve segment contained in the submanifold. We then find a finite cover of this segment by neighborhoods on which ϕ is a local diffeomorphism. Composition of these local maps, together with the group property of ϕ, finally yields the existence of a t such that $\phi(t, \mathcal{A}) = \mathcal{B}$.

Thus, we have finally obtained a parameterization of the intersection submanifold, such that the Hamiltonian flow generated by H separates naturally. In fact, the vector field X_H corresponds to a unit speed flow in $\mathbb{R}^n$ along the t_1 direction with no change in the other coordinates. In fact, since linear combinations of commuting vector fields still commute, we can reparameterize the submanifold so that the flow corresponds to translation in a noncoordinate direction.

In the special case that the intersection submanifold is compact, it is possible to show that there exists a parameterization of the type described above such that $t = \varphi \in \mathbb{T}^n$, and so that the Hamiltonian flow generated by H corresponds to constant frequency angular motion on the torus $\mathbb{T}^n$.

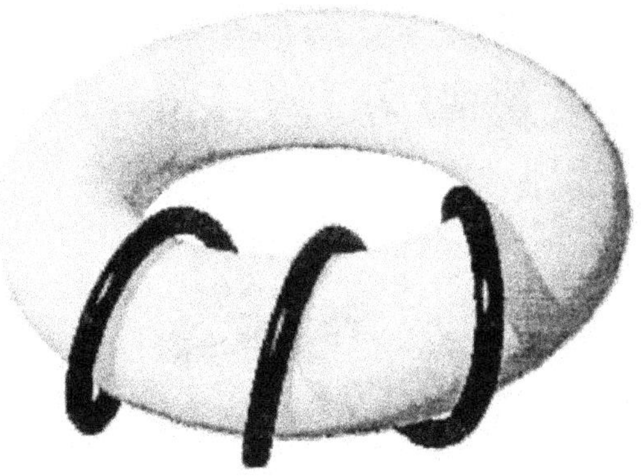

Compact intersections are tori on which orbits are wound.

While the separability of the Hamiltonian flow is a theoretical consequence of the existence of an independent set of n invariants that pairwise Poisson commute, finding the suitable variable transformation is, in practice, not as straightforward. Before we discuss the possibility of achieving separability in a general case, we will describe a special class of transformations with desirable properties.

2.2 Canonical transformations

The analysis of mechanical systems is greatly aided by the existence of a Hamiltonian form of the equations of motion. In particular, it is advantageous to consider parameterizing state space with conjugate coordinates canonically paired as in Eq. (1.61). It is further reasonable to expect that particular forms of the Hamiltonian are preferable to others. For example, we have already noted that the absence of a particular coordinate in the Hamiltonian results in the existence of another conserved quantity, the conjugate momentum. In this section we shall therefore be concerned with finding variable transformations that preserve the canonical pairing while simplifying the Hamiltonian in a desired direction. These transformations are known as canonical or symplectic transformations. In particular, we restrict attention to autonomous Hamiltonians and time-independent transformations since, by the analysis in the previous chapter, nonautonomous Hamiltonian systems are trivially transformed to the former.

2.2.1 *The Poisson bracket conditions*

Consider the coordinate transformation

$$\boldsymbol{q} = \boldsymbol{q}(\bar{\boldsymbol{q}}, \bar{\boldsymbol{p}}) \tag{2.19}$$

and the associated transformation of the conjugate momenta

$$\boldsymbol{p} = \boldsymbol{p}(\bar{\boldsymbol{q}}, \bar{\boldsymbol{p}}) \tag{2.20}$$

from $(\bar{\boldsymbol{q}}, \bar{\boldsymbol{p}}) \in \mathbb{R}^{2m}$ to $(\boldsymbol{q}, \boldsymbol{p}) \in \mathbb{R}^{2n}$, where $m \geq n$. Contrary to standard treatments, we include the possibility of an increase in the number of variables since, as we shall see, this is a central feature of the particular canonical transformation we apply in

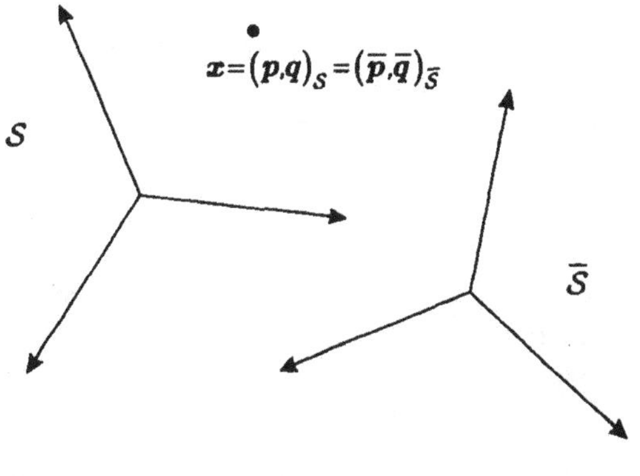

Transforming to a different perspective might simplify the analysis.

later chapters. If $m > n$, the transformation is clearly not one-to-one, but we do require it to be onto.

We now write $\bar{G}(\bar{q},\bar{p})$ for $G(q(\bar{q},\bar{p}),p(\bar{q},\bar{p}))$ for an arbitrary function G, and consider Hamilton's equations in the Poisson bracket form

$$\dot{\bar{F}} = \{\bar{F},\bar{H}\}_{\bar{q},\bar{p}}, \tag{2.21}$$

where the subscript on the bracket denotes the corresponding conjugate variables. Assume that we are given a solution $\bar{q}(t),\bar{p}(t)$ of the Hamiltonian system corresponding to Eq. (2.21). We now wish to inquire into the condition on Eqs. (2.19-2.20) so that the corresponding functions $q(t),p(t)$ are canonically paired through H. In particular, we would like

$$\{F,H\}_{q,p}(t) = \{\bar{F},\bar{H}\}_{\bar{q},\bar{p}}(t), \tag{2.22}$$

since this, together with $\dot{F} = \dot{\bar{F}}$, implies that Hamilton's equations are satisfied in the unbarred variables. However, it is straightforward to show that

$$\begin{aligned}\{\bar{F},\bar{H}\}_{\bar{q},\bar{p}}(t) &= \frac{\partial F}{\partial q_j}(t)\frac{\partial H}{\partial q_k}(t)\{q_j,q_k\}_{\bar{q},\bar{p}}(t) + \frac{\partial F}{\partial p_j}(t)\frac{\partial H}{\partial p_k}(t)\{p_j,p_k\}_{\bar{q},\bar{p}}(t)\\ &+ \left(\frac{\partial F}{\partial q_j}(t)\frac{\partial H}{\partial p_k}(t) - \frac{\partial F}{\partial p_k}(t)\frac{\partial H}{\partial q_j}(t)\right)\{q_j,p_k\}_{\bar{q},\bar{p}}(t)\end{aligned} \tag{2.23}$$

which equals $\{F,H\}_{q,p}(t)$ if the Poisson bracket conditions

$$\{q_j,q_k\}_{\bar{q},\bar{p}}(t) \equiv \{p_j,p_k\}_{\bar{q},\bar{p}}(t) \equiv 0, \quad \{q_j,p_k\}_{\bar{q},\bar{p}}(t) \equiv \delta_{jk}, \ \forall j,k \tag{2.24}$$

hold for the given orbit. Here, δ_{jk} denotes the Kronecker delta. That these conditions are, in general, also necessary follows by considering $F = q_i$ or $F = p_i$.

We note that the Poisson bracket expressions (2.24) provide conditions on arbitrary orbits in the barred variables. Now consider the situation where there exists an invariant submanifold in the barred state space such that (2.24) are satisfied. Then, provided the restriction of the transformation to the submanifold is still onto, the transformation is canonical and we are free to consider the dynamical system given by the transformed Hamiltonian on the submanifold.

2.2.2 *Mixed variable generating functions*

The method of generating functions offers a convenient way of automatically producing canonical transformations independently of the particular Hamiltonian system at hand. From the previous chapter we know that Hamilton's equations are equivalent to the modified Hamilton's principle

$$\delta\int_{t_1}^{t_2} (p\cdot\dot{q} - H(p,q))\,dt = 0, \tag{2.25}$$

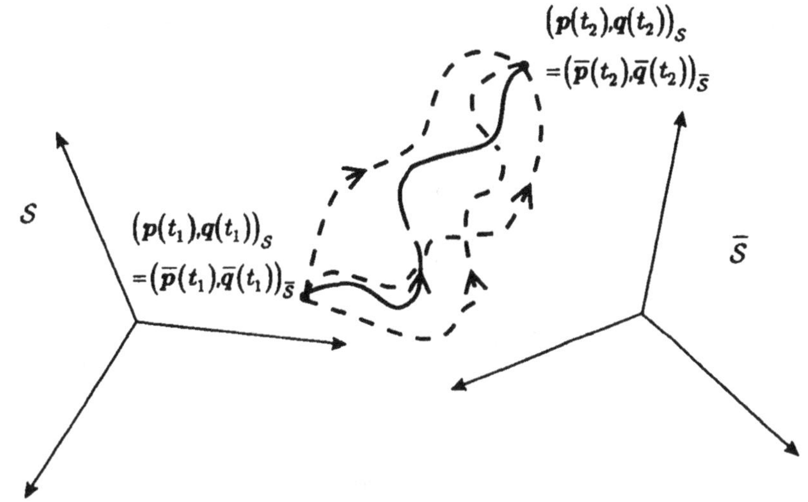

The particular curve satisfying the modified Hamilton principle is independent of the coordinate system.

which should hold in any set of conjugate variables. It thus follows that the integrands expressed in two different sets of conjugate variables can differ only by a total time-derivative of a function S of the old and new variables (and generally time, although we restrict ourselves to time-independent functions)

$$\boldsymbol{p}\cdot\dot{\boldsymbol{q}} - H(\boldsymbol{p},\boldsymbol{q}) = \bar{\boldsymbol{p}}\cdot\dot{\bar{\boldsymbol{q}}} - \bar{H}(\bar{\boldsymbol{p}},\bar{\boldsymbol{q}}) + \frac{dS}{dt}(\boldsymbol{q},\boldsymbol{p},\bar{\boldsymbol{q}},\bar{\boldsymbol{p}}) \tag{2.26}$$

where

$$\frac{dS}{dt} = \frac{\partial S}{\partial \boldsymbol{q}}\cdot\dot{\boldsymbol{q}} + \frac{\partial S}{\partial \boldsymbol{p}}\cdot\dot{\boldsymbol{p}} + \frac{\partial S}{\partial \bar{\boldsymbol{q}}}\cdot\dot{\bar{\boldsymbol{q}}} + \frac{\partial S}{\partial \bar{\boldsymbol{q}}}\cdot\dot{\bar{\boldsymbol{q}}}. \tag{2.27}$$

Given a canonical variable transformation as in Eqs. (2.19-2.20), we now wish to find conditions on S for Eq. (2.26) to hold independently of the dynamics induced by the corresponding Hamiltonian system. Since the transformation is canonical, $H = \bar{H}$. Substituting Eqs. (2.19-2.20,2.27) into Eq. (2.26) and collecting coefficients, we find

$$\left[p_i - \bar{p}_j\frac{\partial \bar{q}_j}{\partial q_i} - \frac{\partial S}{\partial q_i} - \frac{\partial S}{\partial \bar{q}_j}\frac{\partial \bar{q}_j}{\partial q_i} - \frac{\partial S}{\partial \bar{p}_j}\frac{\partial \bar{p}_j}{\partial q_i}\right]\dot{q}_i$$
$$-\left[\bar{p}_j\frac{\partial \bar{q}_j}{\partial p_i} + \frac{\partial S}{\partial p_i} + \frac{\partial S}{\partial \bar{q}_j}\frac{\partial \bar{q}_j}{\partial p_i} + \frac{\partial S}{\partial \bar{p}_j}\frac{\partial \bar{p}_j}{\partial p_i}\right]\dot{p}_i = 0. \tag{2.28}$$

Since we require that the above be an identity regardless of the dynamics of the Hamiltonian system, it follows that the coefficients of $\dot{q}_i, \dot{p}_i$ must all vanish. For example, for the identity transformation we find $S \equiv 0$.

In fact, nontrivial transformations can be generated using Eq. (2.28). For example, Eq. (2.28) is satisfied by $S = S(\boldsymbol{q}, \bar{\boldsymbol{q}})$ such that

$$\boldsymbol{p} = \frac{\partial S}{\partial \boldsymbol{q}}, \qquad \bar{\boldsymbol{p}} = -\frac{\partial S}{\partial \bar{\boldsymbol{q}}}. \tag{2.29}$$

Similarly, $S = S'(\bar{\boldsymbol{q}}, \boldsymbol{p}) + \boldsymbol{p} \cdot \boldsymbol{q}$ such that

$$\bar{\boldsymbol{p}} = -\frac{\partial S'}{\partial \bar{\boldsymbol{q}}}, \qquad \boldsymbol{q} = -\frac{\partial S'}{\partial \boldsymbol{p}} \tag{2.30}$$

and $S = S'(\boldsymbol{q}, \bar{\boldsymbol{p}}) - \bar{\boldsymbol{p}} \cdot \bar{\boldsymbol{q}}$ such that

$$\boldsymbol{p} = \frac{\partial S'}{\partial \boldsymbol{q}}, \qquad \bar{\boldsymbol{q}} = \frac{\partial S'}{\partial \bar{\boldsymbol{p}}} \tag{2.31}$$

are solutions to Eq. (2.28). Reverting this procedure we see that for an arbitrary function S or S', Eqs. (2.29-2.31) generate canonical transformations independently of the Hamiltonian structure at hand. We call functions like S and S' above **mixed variable generating functions.** Characteristically they depend on one half of the unbarred conjugate variables and one half of the barred conjugate variables. We shall find particular use for $S'(\bar{\boldsymbol{q}}, \boldsymbol{p})$ and $S'(\boldsymbol{q}, \bar{\boldsymbol{p}})$, and will henceforth dispense with the prime notation.

The canonical transformation generated by $S(\bar{\boldsymbol{q}}, \boldsymbol{p})$ is locally invertible provided that we can invert the first of Eqs. (2.30). The requirement on S is then that the Jacobian

$$\det \left(\frac{\partial^2 S}{\partial \bar{\boldsymbol{q}} \partial \boldsymbol{p}} \right) \tag{2.32}$$

be nonzero.

As a first example of generating functions of the above type, we consider those linear in the old momenta $\boldsymbol{p}$:

$$S(\bar{\boldsymbol{q}}, \boldsymbol{p}) = -\boldsymbol{p} \cdot \boldsymbol{q}(\bar{\boldsymbol{q}}). \tag{2.33}$$

From Eqs. (2.30) it now follows that

$$\boldsymbol{q} = \boldsymbol{q}(\bar{\boldsymbol{q}}) \tag{2.34}$$

and the corresponding momentum transformation

$$\bar{p}_i = \boldsymbol{p} \cdot \frac{\partial \boldsymbol{q}}{\partial \bar{q}_i}, i = 1, \ldots, n. \tag{2.35}$$

Thus, we see how a transformation between the old and new position coordinates, not explicitly involving the conjugate momenta, can be conveniently expressed in the mixed variable generating function formalism. This is how Eqs. (1.100-1.101) were obtained. The invertibility condition (2.32) now becomes

$$\det\left(\frac{\partial \boldsymbol{q}}{\partial \bar{\boldsymbol{q}}}\right) \neq 0, \tag{2.36}$$

which is equivalent to local invertibility of the coordinate transformation alone. Similarly, transformations of the momenta not involving the positions can be generated by generating functions linear in the new position coordinates.

Illustration

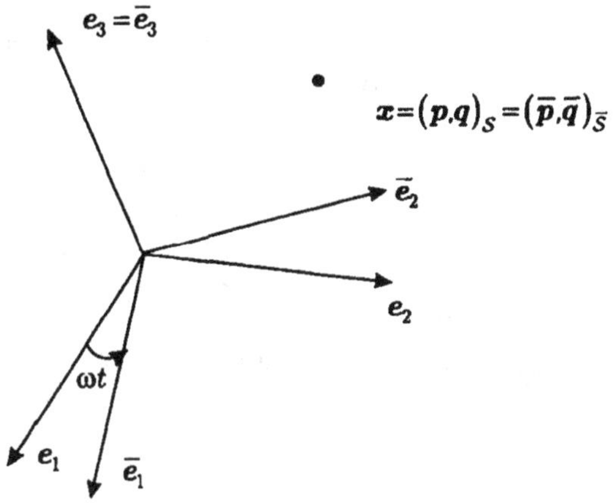

A change to a rotating reference frame.

We consider the transformation from an inertial reference frame to one rotating with a constant rotation rate, ω, such that the two coordinate systems have coinciding origins and z-axes and, furthermore, such that their axes coincide at time 0. The corresponding generating function is then

$$S_{rot}(\bar{q}_0, \bar{\boldsymbol{q}}, p_0, \boldsymbol{p}) = -p_0\bar{q}_0 - p_1(\bar{q}_1 \cos\omega\bar{q}_0 - \bar{q}_2 \sin\omega\bar{q}_0)$$

$$-p_2(\bar{q}_1 \sin\omega\bar{q}_0 + \bar{q}_2 \cos\omega\bar{q}_0) - p_3\bar{q}_3. \tag{2.37}$$

Note that we have here assumed that a suspended formalism has been adopted such that the zero subscript coordinate is equivalent to time. The corresponding coordinate transformation then becomes

$$q_0 = \bar{q}_0, \quad q_1 = \bar{q}_1 \cos\omega\bar{q}_0 - \bar{q}_2 \sin\omega\bar{q}_0,$$

$$q_2 = \bar{q}_1 \sin\omega\bar{q}_0 + \bar{q}_2 \cos\omega\bar{q}_0, \quad q_3 = \bar{q}_3 \tag{2.38}$$

and the determinant (2.36) equals 1 thus guaranteeing local invertibility. In fact, the momentum transformation can be written

$$p_0 = \bar{p}_0, \quad p_1 = \bar{p}_1 \cos\omega q_0 - \bar{p}_2 \sin\omega q_0,$$

$$p_2 = \bar{p}_1 \sin\omega q_0 + \bar{p}_2 \cos\omega q_0, \quad p_3 = \bar{p}_3. \tag{2.39}$$

Transformations of this type will reappear in later chapters when physical applications are considered. •

We shortly describe how generating functions in principle allow for the full determination of the Hamiltonian flow. But there still remains to discuss two important transformations that require slightly different considerations.

2.2.3 Scaling the independent and dependent variables

We return to the integral in the modified Hamilton's principle

$$\int_{t_1}^{t_2} (\boldsymbol{p}\cdot\dot{\boldsymbol{q}} - H(\boldsymbol{p},\boldsymbol{q}))\,\mathrm{d}t = 0 \tag{2.40}$$

and consider the effect of transformations of the independent variable of the form

$$\frac{d}{ds} = f(\boldsymbol{q},\boldsymbol{p})\frac{d}{dt}, \tag{2.41}$$

where $f > 0$. Introducing this transformation in Eq. (2.40) we obtain

$$\int_{s_1}^{s_2} (\boldsymbol{p}\cdot\boldsymbol{q}' - f(\boldsymbol{q},\boldsymbol{p})H(\boldsymbol{p},\boldsymbol{q}))\,\mathrm{d}s = 0 \tag{2.42}$$

where $'$ denotes differentiation with respect to s. If we consider applying the variation implied in the modified Hamilton's principle over a fixed interval in s, the canonical pairing of the state variables through Hamilton's equations is preserved if we adopt

$$H_{new}(\boldsymbol{p},\boldsymbol{q}) = f(\boldsymbol{q},\boldsymbol{p})H(\boldsymbol{p},\boldsymbol{q}) \tag{2.43}$$

as our new Hamiltonian. Consider $\boldsymbol{q}(s),\boldsymbol{p}(s)$ as a solution to Hamilton's equation with Hamiltonian H_{new}. From Eq. (2.41) we then obtain $t(s)$, which can be inverted to yield $\boldsymbol{q}(t) = \boldsymbol{q}(s(t))$ and similarly for $\boldsymbol{p}$. Now,

$$\dot{\boldsymbol{q}} = \frac{1}{f}\boldsymbol{q}' = \frac{1}{f}\frac{\partial H_{new}}{\partial \boldsymbol{p}} = \frac{\partial H}{\partial \boldsymbol{p}} + \frac{1}{f}\frac{\partial f}{\partial \boldsymbol{p}}H \tag{2.44}$$

and

$$\dot{\boldsymbol{p}} = \frac{1}{f}\boldsymbol{p}' = -\frac{1}{f}\frac{\partial H_{new}}{\partial \boldsymbol{q}} = -\frac{\partial H}{\partial \boldsymbol{q}} - \frac{1}{f}\frac{\partial f}{\partial \boldsymbol{q}}H. \tag{2.45}$$

In order for $\boldsymbol{q}$ and $\boldsymbol{p}$ to be canonically paired with Hamiltonian H and time t, these equations must reduce to $\frac{\partial H}{\partial \boldsymbol{p}}$ and $-\frac{\partial H}{\partial \boldsymbol{q}}$, respectively. For general f this requires $H \equiv 0$. However, we see that H_{new} is conserved under the motion $\boldsymbol{q}(s)$ and $\boldsymbol{p}(s)$ and $H \equiv 0$ is thus guaranteed if we choose initial conditions in the s variable such that

$$H_{new}(s = 0) = 0. \tag{2.46}$$

Thus, we have shown that solutions to the Hamiltonian system generated by H_{new}, with H_{new} initially vanishing, correspond to motion in the original Hamiltonian system on the zero energy submanifold under the time transformation (2.41). But we are assuming that the suspended formalism has been adopted so that, in fact, all orbits in the original unsuspended Hamiltonian system are recovered in this way.

Illustration

We consider the Hamiltonian in Eq. (1.99) for the perturbed two-body problem

$$H(\boldsymbol{p}, \boldsymbol{r}, t) = \frac{1}{2}\boldsymbol{p}\cdot\boldsymbol{p} - \frac{\mu}{r} + V(\boldsymbol{r}, t). \tag{2.47}$$

We would like to eliminate the r^{-1} singularity in the Hamiltonian. From the above discussion this could be effected by the time transformation

$$\frac{dt}{ds} = r. \tag{2.48}$$

Before applying this transformation, we need to introduce t as a new coordinate. The suspended formalism then allows us to consider the suspended Hamiltonian

$$H(p_0, \boldsymbol{p}, x_0, \boldsymbol{r}) = \frac{1}{2}\boldsymbol{p}\cdot\boldsymbol{p} - \frac{\mu}{r} + V(\boldsymbol{r}, x_0) + p_0 \tag{2.49}$$

and finally obtain

$$H_{new}(p_0, \boldsymbol{p}, x_0, \boldsymbol{r}) = \frac{r}{2}\boldsymbol{p}\cdot\boldsymbol{p} - \mu + rV(\boldsymbol{r}, x_0) + rp_0, \tag{2.50}$$

which is regular, provided no additional singularities are present in the potential V. •

Finally, we consider transformations of the type

$$\boldsymbol{q} = \alpha\bar{\boldsymbol{q}}, \quad \boldsymbol{p} = \beta\bar{\boldsymbol{p}} \tag{2.51}$$

where α and β are constants. While this is not a canonical transformation, the barred variables are still canonically paired through the Hamiltonian

$$\bar{H}(\bar{\boldsymbol{q}},\bar{\boldsymbol{p}}) = \frac{1}{\alpha\beta}H(\alpha\bar{\boldsymbol{q}},\beta\bar{\boldsymbol{p}}). \tag{2.52}$$

This is obviously convenient in reducing the number of important physical parameters in a given problem.

2.3 Integrability versus nonintegrability

In the first section of this chapter we indicated how an n-degree-of-freedom Hamiltonian flow can be completely separated if n functionally independent, pairwise Poisson commuting invariants of the motion are given. We showed how the intersections of the invariant submanifolds corresponding to level surfaces of the invariants could be parameterized in a natural way so that the flow appeared as independent constant flow in n coordinates. In particular, when the intersection was compact, this parameterization could be achieved with the coordinates being angles and the motion submanifold equivalent to an n-dimensional torus. Simply put, the Hamiltonian flow could then be described by the differential equations

$$\dot{\boldsymbol{F}} = 0, \dot{\boldsymbol{\varphi}} = \boldsymbol{\omega} \tag{2.53}$$

where $\varphi_i \in \mathbb{S}^1$ are angle variables and ω_i constant frequencies.

It is not generally the case that $\boldsymbol{F}$ and $\boldsymbol{\varphi}$ are canonically paired. Instead, there exists a function $\boldsymbol{I}(\boldsymbol{F})$ such that the set of I_i (known as action variables) consists of conjugate momenta to the φ_i. It follows that

$$\dot{\boldsymbol{I}} = 0, \tag{2.54}$$

i.e., the Hamiltonian in terms of $\boldsymbol{I}$ and $\boldsymbol{\varphi}$ is independent of the angle variables. Moreover,

$$\{I_i, I_j\} = \frac{\partial I_i}{\partial F_k}\frac{\partial I_j}{\partial F_l}\{F_k, F_l\} = 0; \tag{2.55}$$

i.e., the action variables Poisson commute. This also follows from their conjugate character to the angle coordinates.

We shall say that a system that is separable in the sense described above is **integrable**. Thus, integrability relies on the existence of n invariants with the properties described previously or, equivalently, a transformation to new canonical state variables for which the new Hamiltonian only depends on the momenta. For

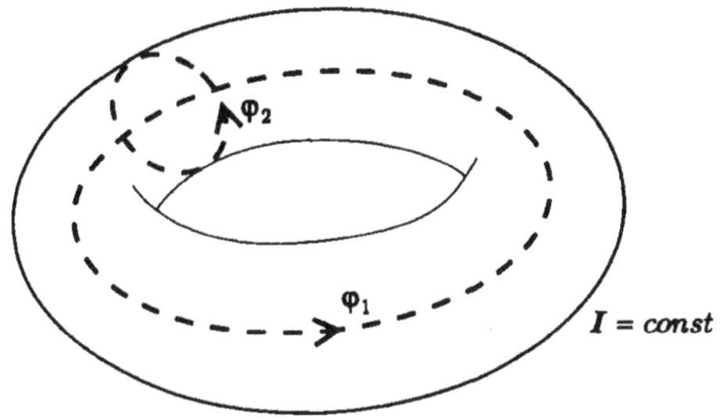

The motion along the torus preserves the value of I, while increasing φ at a constant rate.

a general Hamiltonian system, the integrability requirement thus translates to the existence of a generating function, $S(\boldsymbol{q}, \bar{\boldsymbol{p}})$ such that

$$\bar{H}(\bar{\boldsymbol{q}}, \bar{\boldsymbol{p}}) = H(\boldsymbol{q}(\bar{\boldsymbol{q}}, \bar{\boldsymbol{p}}), \boldsymbol{p}(\bar{\boldsymbol{q}}, \bar{\boldsymbol{p}})) = H(\boldsymbol{q}, \frac{\partial S}{\partial \boldsymbol{q}}) = \bar{H}(\bar{\boldsymbol{p}}) \tag{2.56}$$

where the first equality is the definition of $\bar{H}$, the second follows from Eq. (2.31), and the last is a consequence of integrability. Note, however, that all barred coordinates are cyclic and, hence, the momenta are all invariants. In other words, the right-hand side is a constant for each orbit. Let us then consider $\bar{\boldsymbol{p}} = \boldsymbol{\alpha}$ as a constant parameter and the problem becomes finding a function $S(\boldsymbol{q}; \boldsymbol{\alpha})$ such that

$$H(\boldsymbol{q}, \frac{\partial S}{\partial \boldsymbol{q}}) = \Psi(\boldsymbol{\alpha}), \tag{2.57}$$

where Ψ is an arbitrary function of its arguments. This partial differential equation is known as **Hamilton-Jacobi's equation**. As discussed above, its solvability guarantees integrability and vice versa.

Illustration

The class of Liouville problems consists of all systems whose Hamiltonian separates as

$$H(\boldsymbol{q}, \boldsymbol{p}) = \frac{1}{2} \frac{\sum_{i=1}^{n} (p_i^2/m_i(q_i) + V_i(q_i))}{\sum_{i=1}^{n} f_i(q_i)}, \tag{2.58}$$

where the denominator is strictly greater than zero. With the desire to remove the $\left(\sum_{i=1}^{n} f_i(q_i)\right)^{-1}$ singularity, we introduce time as position coor-

dinate, as described in previous sections. Finally, the time transformation

$$\frac{dt}{ds} = \sum_{i=1}^{n} f_i(q_i) \tag{2.59}$$

yields the Hamiltonian

$$H(\boldsymbol{q}, p_0, \boldsymbol{p}) = \frac{1}{2}\sum_{i=1}^{n} \left(p_i^2/m_i(q_i) + V_i(q_i) + p_0 f_i(q_i)\right). \tag{2.60}$$

Hamilton-Jacobi's equation then becomes

$$\frac{1}{2}\sum_{i=1}^{n}\left(\left(\frac{\partial S}{\partial q_i}\right)^2 /m_i(q_i) + V_i(q_i) + \frac{\partial S}{\partial q_0} f_i(q_i)\right) = \Psi(\boldsymbol{\alpha}). \tag{2.61}$$

We now assume a separable solution $S(\boldsymbol{q}) = \sum_{i=0}^{n} S_i(q^i)$ *and note that* $\frac{\partial S_0}{\partial q_0}$ *is the only* q_0 *dependent term in (2.61) and must thus be constant. We define this constant to be* α_0. *Furthermore, we consider* $\Psi = \sum_{i=1}^{n} \alpha_i$ *such that*

$$\left(\frac{\partial S}{\partial q_i}\right)^2 /m_i(q_i) + V_i(q_i) + \alpha_0 f_i(q_i) = 2\alpha_i,\ i = 1, \ldots, n. \tag{2.62}$$

This yields

$$\frac{\partial S}{\partial q_i} = \sqrt{m_i(q_i)}\sqrt{2\alpha_i - V_i(q_i) - \alpha_0 f_i(q_i)},\ i = 1, \ldots, n, \tag{2.63}$$

which can be solved by quadrature. The transformed Hamiltonian is

$$\bar{H}(\bar{\boldsymbol{p}}) = \sum_{i=1}^{n} \bar{p}_i \tag{2.64}$$

and the solution simply

$$\bar{\boldsymbol{p}} \equiv \bar{\boldsymbol{p}}(s=0), \text{ and } \bar{\boldsymbol{q}} = s\boldsymbol{e} + \bar{\boldsymbol{q}}(s=0), \tag{2.65}$$

where $e_i = 1$, $i = 1, \ldots, n$. *Thus, Liouville problems are integrable.* •

In the illustration we saw how a separable solution was possible, provided the Hamiltonian was of suitable form. In particular, we saw that

$$\frac{\partial S}{\partial q_i} = g_i(q_i, \bar{p}_0, \bar{p}_i), \text{ (no sum)}, \tag{2.66}$$

which from Eq. (2.31) also constitutes the equation giving the old momenta in terms of the new. The corresponding coordinate transformation is obtained from

$$\bar{q}_i = \frac{\partial S}{\partial \bar{p}_i}, \tag{2.67}$$

which needs to be inverted with respect to the old coordinates to complete the canonical transformation. Invertibility is guaranteed provided that the Jacobian

$$\frac{\partial \bar{q}_i}{\partial q_j} = \frac{\partial^2 S}{\partial q_j \partial \bar{p}_i} = \frac{\partial g_j}{\partial \bar{p}_i} \tag{2.68}$$

is nonzero. Since g_j depends only on $\bar{p}_0$ and $\bar{p}_j$, the Jacobian is triangular and invertibility thus follows from the nonvanishing of each of the factors $\partial g_i/\partial \bar{p}_i$.

It turns out that only very few physically motivated Hamiltonian systems admit a global solution to Hamilton-Jacobi's equation. These are, in general, separable in some set of variables as a consequence of certain physical symmetries. In fact, it is possible to show that a large class of Hamiltonian systems are **nonintegrable**; i.e., no set of n global invariants with the above properties can be found. Instead, the motion is restricted to submanifolds of dimension higher than n and not reducible to an n-dimensional manifold. Such motions are much more complicated than the torus motion described above. Indeed, very complex stochastic behavior is observed of the type discussed in Chapter 3. While, in general, no global set of invariants can be found, it is still possible that a complete set of local invariants may be found. In this case, the intersection of their level sets is again n-dimensional, and if it is compact the motion is that on a torus.

A detail from figure in Chapter 1. The closed curves correspond to intersections of n-dimensional tori with the plane (q_2, p_2). On the contrary, for the randomly distributed points in between the tori, no complete set of invariants exists.

In the next section we discuss the case of small perturbations from an integrable system and indicate how one may prove that, in fact, a majority of unperturbed tori survive the perturbation, although the Hamiltonian flow is globally nonintegrable.

2.4 The KAM theorem

2.4.1 *Formal integrability*

In this section we consider nearly-integrable Hamiltonian systems of the form

$$H_\varepsilon(\boldsymbol{q},\boldsymbol{p};\varepsilon) = H_0(\boldsymbol{q},\boldsymbol{p}) + \varepsilon H_1(\boldsymbol{q},\boldsymbol{p}), \tag{2.69}$$

such that the unperturbed Hamiltonian H_0 admits a global solution to Hamilton-Jacobi's equation. In particular, we assume the existence of invariant tori for the unperturbed flow such that a suitable canonical transformation yields

$$H_\varepsilon(\boldsymbol{I},\varphi;\varepsilon) = H_0(\boldsymbol{I}) + \varepsilon H_1(\boldsymbol{I},\varphi), \tag{2.70}$$

where H_1 is periodic in the angle variables. If the perturbed Hamiltonian were integrable, there should exist a canonical transformation given from the mixed variable generating function $S(\bar{\boldsymbol{I}},\varphi)$ such that

$$H_0(\frac{\partial S}{\partial \varphi}) + \varepsilon H_1(\frac{\partial S}{\partial \varphi},\varphi) = \Psi(\bar{\boldsymbol{I}}) \tag{2.71}$$

for all $\bar{\boldsymbol{I}}$. The form of the perturbed Hamiltonian suggests a power series expansion ansatz for S:

$$S = \sum_{k=0}^{\infty} \varepsilon^k S_k(\bar{\boldsymbol{I}},\varphi) \tag{2.72}$$

and similarly for Ψ. Substituting this expression into Eq. (2.71) and letting $\varepsilon = 0$ yields

$$H_0(\frac{\partial S_0}{\partial \varphi}) = \Psi_0, \tag{2.73}$$

where the right-hand side is independent of φ. A possible solution to this equation is obtained from

$$S_0(\bar{\boldsymbol{I}},\varphi) = \bar{\boldsymbol{I}} \cdot \varphi \tag{2.74}$$

and consequently,

$$\Psi_0(\bar{\boldsymbol{I}}) = H_0(\bar{\boldsymbol{I}}). \tag{2.75}$$

Identifying the coefficients of ε^k for $k \geq 1$ yields

$$\omega(\bar{\boldsymbol{I}}) \cdot \frac{\partial S_k}{\partial \varphi}(\bar{\boldsymbol{I}},\varphi) + N_k(\bar{\boldsymbol{I}},\varphi) = \Psi_k(\bar{\boldsymbol{I}}), \tag{2.76}$$

where

$$\omega(\bar{\boldsymbol{I}}) = \frac{\partial H_0}{\partial \boldsymbol{I}}(\bar{\boldsymbol{I}}) \tag{2.77}$$

is the vector of frequencies on the unperturbed tori. Also, N_k is a φ dependent term that involves only S_i, $i < k$. We can thus solve Eq. (2.76) iteratively, at each step faced with a linear partial differential equation in S_k. The periodicity of N_k suggests expanding all φ dependent quantities in Fourier series

$$N_k(\varphi) = \sum_{m} N_{k,m} e^{im\cdot\varphi}, \text{ and } S_k(\varphi) = \sum_{m} S_{k,m} e^{im\cdot\varphi} \tag{2.78}$$

where we have omitted the $\bar{I}$ dependence and the summation is over all integer vectors, m. Eq. (2.76) then becomes

$$\mathrm{i}\sum_{m\neq 0} \omega \cdot m S_{k,m} e^{im\cdot\varphi} + \sum_{m} N_{k,m} e^{im\cdot\varphi} = \Psi_k. \tag{2.79}$$

Identifying different Fourier coefficients then yields

$$\Psi_k(\bar{I}) = N_{k,0}(\bar{I}), \tag{2.80}$$

$$S_{k,m}(\bar{I}) = \mathrm{i}\frac{N_{k,m}(\bar{I})}{\omega(\bar{I})\cdot m}, \; m \neq 0, \tag{2.81}$$

and $S_{k,0} = 0$. Finally, we obtain the generating function

$$S(\bar{I},\varphi) = \bar{I}\cdot\varphi + \mathrm{i}\sum_{k=1}^{\infty} \varepsilon^k \sum_{m\neq 0} \frac{N_{k,m}(\bar{I})}{\omega(\bar{I})\cdot m} e^{im\cdot\varphi} \tag{2.82}$$

If we could prove the global convergence of this series, our task would be complete. Indeed, we would then have found a global canonical transformation to a set of action-angle coordinates, and the flow would thus be integrable. However, already the summation over m is likely to diverge if $\omega\cdot m$ is close to zero for infinitely many m. If we write

$$\omega\cdot m = \omega_1(m_1 + m_2\frac{\omega_2}{\omega_1} + \ldots + m_n\frac{\omega_n}{\omega_1}) \tag{2.83}$$

and consider approximating the ratios $\frac{\omega_k}{\omega_1}$ increasingly more closely by rational numbers, we can find an infinite sequence of integer vectors m such that the right-hand side is arbitrarily close to zero. Clearly, in the case that there exist rational relations between the unperturbed frequencies, the denominator vanishes and the entire expansion scheme fails unless the corresponding numerators also vanish. Hence, we shall primarily consider the absence of rational relations, or unperturbed flows with noncommensurable frequencies.

Illustration

From the generating function S we obtain the canonical transformation $\varphi(\bar{\varphi}, \bar{\boldsymbol{I}})$ and $\boldsymbol{I}(\bar{\varphi}, \bar{\boldsymbol{I}})$ and, solving for the motion in the barred variables, we then obtain the time-evolution of the original state variables. The Hamiltonian in the barred variables is simply $\Psi(\bar{\boldsymbol{I}})$ and thus, $\bar{\boldsymbol{I}}$ is constant and the frequency in $\bar{\varphi}$ variables,

$$\bar{\omega}(\bar{\boldsymbol{I}}) = \frac{\partial \Psi}{\partial \bar{\boldsymbol{I}}}(\bar{\boldsymbol{I}}) = \omega(\bar{\boldsymbol{I}}) + \mathcal{O}(\varepsilon), \tag{2.84}$$

where the last equality follows from Eq. (2.75). A natural question is now whether a particular unperturbed torus is preserved; i.e., if the frequencies of the motion on the perturbed torus are the same as for the original torus motion, $\omega(\boldsymbol{I}_0) = \omega_0$. From Eq. (2.84) this amounts to finding $\bar{\boldsymbol{I}}(\varepsilon)$ such that

$$\omega_0 = \omega(\bar{\boldsymbol{I}}) + \mathcal{O}(\varepsilon). \tag{2.85}$$

The implicit function theorem implies that a solution to this equation exists for sufficiently small ε provided that the Jacobian

$$\det\left(\frac{\partial \omega}{\partial \boldsymbol{I}}\right)(\boldsymbol{I}_0) = \det\left(\frac{\partial^2 H_0}{\partial \boldsymbol{I}^2}\right)(\boldsymbol{I}_0) \tag{2.86}$$

is nonzero. This nondegeneracy condition thus assures the formal existence of quasiperiodic power series expansions in ε with $\bar{\omega} = \omega_0$ for sufficiently small ε. •

Although we were able to formally solve the Hamilton-Jacobi's equation for the appropriate generating function, we have found that, unless the $N_{k,n}$ fall off sufficiently fast, the series generally diverges. The question arises whether divergence is the only possibility or if, in fact, there exist choices for $\bar{\boldsymbol{I}}$ for which convergence can be obtained. This is indeed the assertion of the Kolmogorov-Arnold-Moser (KAM) theory named after its originators.

2.4.2 Questions of convergence

In this section we revisit the series solution (2.82) so as to better appreciate the source of its divergence. It is advantageous to expand H_0 and H_1 in Taylor expansions around $\bar{\boldsymbol{I}}$:

$$H_i(\varphi, \frac{\partial S}{\partial \varphi}) = \sum_{j \geq 0} H_{i,j}(\varphi, \bar{\boldsymbol{I}})(\frac{\partial \tilde{S}}{\partial \varphi}, \ldots, \frac{\partial \tilde{S}}{\partial \varphi}), \; i = 0, 1 \tag{2.87}$$

where $H_{i,j}(\varphi, \bar{\boldsymbol{I}})$ is a multilinear function of j vectors and $\tilde{S} = S - \bar{\boldsymbol{I}} \cdot \varphi$. For example,

$$H_{0,1}(\bar{\boldsymbol{I}})(\frac{\partial \tilde{S}}{\partial \varphi}) = \frac{\partial H_0}{\partial \boldsymbol{I}}(\bar{\boldsymbol{I}}) \cdot \frac{\partial \tilde{S}}{\partial \varphi} = \omega(\bar{\boldsymbol{I}}) \cdot \frac{\partial \tilde{S}}{\partial \varphi}. \tag{2.88}$$

In this notation, Hamilton-Jacobi's equation may be written

$$\sum_{j \neq 1} H_{0,j}(\bar{\boldsymbol{I}})(\frac{\partial \tilde{S}}{\partial \varphi}, \dots, \frac{\partial \tilde{S}}{\partial \varphi}) + \varepsilon \sum_{k \geq 0} H_{1,k}(\varphi, \bar{\boldsymbol{I}})(\frac{\partial \tilde{S}}{\partial \varphi}, \dots, \frac{\partial \tilde{S}}{\partial \varphi}) = -H_{0,1}(\bar{\boldsymbol{I}})(\frac{\partial \tilde{S}}{\partial \varphi}) + \Psi(\bar{\boldsymbol{I}}). \tag{2.89}$$

To a given order in ε we see that the left-hand side contains only terms corresponding to lower orders in ε and, hence, known quantities. The right-hand side, however, is unknown and the equation is thus aptly suited for iteration. As in the previous section, we assume the series expansion

$$\tilde{S} = \sum_{k \geq 1} \varepsilon^k S_k \tag{2.90}$$

and further expand each S_k in a Fourier series in φ as in Eq. (2.78). Before giving an explicit method for writing down an arbitrary term in this Fourier series in terms of H_0 and H_1, we consider Eq. (2.89) and identify coefficients of ε^1 and ε^2, respectively.

Using the linearity of $H_{i,j}$, we find to first order in ε

$$\sum_{\boldsymbol{m}} H_{1,0,\boldsymbol{m}} e^{\mathrm{i}\boldsymbol{m} \cdot \varphi} = H_{1,0}(\varphi) = -H_{0,1}(\frac{\partial S_1}{\partial \varphi}) + \Psi_1 = -\mathrm{i} \sum_{\boldsymbol{m} \neq 0} S_{1,\boldsymbol{m}} H_{0,1}(\boldsymbol{m}) e^{\mathrm{i}\boldsymbol{m} \cdot \varphi} + \Psi_1 \tag{2.91}$$

where we have omitted explicit inclusion of the $\bar{\boldsymbol{I}}$ argument. Projecting onto each Fourier mode then gives

$$S_{1,\boldsymbol{m}} = \mathrm{i} \frac{H_{1,0,\boldsymbol{m}}}{H_{0,1}(\boldsymbol{m})}, \quad \boldsymbol{m} \neq \boldsymbol{0} \text{ and } \Psi_1 = H_{1,0,0}. \tag{2.92}$$

To second order in ε we obtain

$$H_{0,2}(\frac{\partial S_1}{\partial \varphi}, \frac{\partial S_1}{\partial \varphi}) + H_{1,1}(\varphi)(\frac{\partial S_1}{\partial \varphi}) = -H_{0,1}(\frac{\partial S_2}{\partial \varphi}) + \Psi_2 = -\mathrm{i} \sum_{\boldsymbol{m} \neq 0} S_{2,\boldsymbol{m}} H_{0,1}(\boldsymbol{m}) e^{\mathrm{i}\boldsymbol{m} \cdot \varphi} + \Psi_2. \tag{2.93}$$

We rewrite the left-hand side as

$$(\mathrm{i})^4 \sum_{\boldsymbol{m}_1, \boldsymbol{m}_2 \neq 0} \frac{H_{0,2}(H_{1,0,\boldsymbol{m}_1} \boldsymbol{m}_1, H_{1,0,\boldsymbol{m}_2} \boldsymbol{m}_2)}{H_{0,1}(\boldsymbol{m}_1) H_{0,1}(\boldsymbol{m}_2)} e^{\mathrm{i}(\boldsymbol{m}_1 + \boldsymbol{m}_2) \cdot \varphi}$$
$$+ (\mathrm{i})^2 \sum_{\boldsymbol{m}_1 \neq 0, \boldsymbol{m}_2} \frac{H_{1,1,\boldsymbol{m}_2}(H_{1,0,\boldsymbol{m}_1} \boldsymbol{m}_1)}{H_{0,1}(\boldsymbol{m}_1)} e^{\mathrm{i}(\boldsymbol{m}_1 + \boldsymbol{m}_2) \cdot \varphi}. \tag{2.94}$$

Identifying the Fourier coefficients then yields

$$\Psi_2 = \sum_{m_1=-m_2\neq 0} (i)^4 \frac{H_{0,2}(H_{1,0,m_1} m_1, H_{1,0,m_2} m_2)}{H_{0,1}(m_1) H_{0,1}(m_2)} + (i)^2 \frac{H_{1,1,m_2}(H_{1,0,m_1} m_1)}{H_{0,1}(m_1)} \tag{2.95}$$

and

$$S_{2,m} = \frac{i}{H_{0,1}(m)} \left[\sum_{m_1,m_2\neq 0, m_1+m_2=m} (i)^4 \frac{H_{0,2}(H_{1,0,m_1} m_1, H_{1,0,m_2} m_2)}{H_{0,1}(m_1) H_{0,1}(m_2)} \right.$$

$$\left. + \sum_{m_1\neq 0, m_2, m_1+m_2=m} (i)^2 \frac{H_{1,1,m_2}(H_{1,0,m_1} m_1)}{H_{0,1}(m_1)} \right], \; m \neq 0. \tag{2.96}$$

Illustration

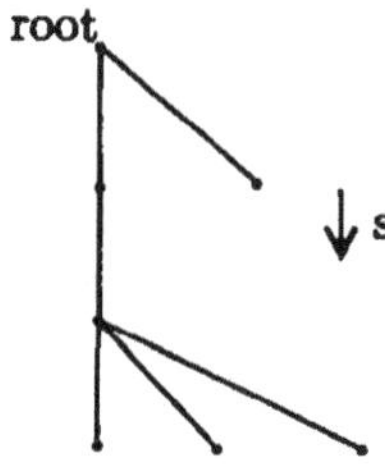

The typical tree structure.

Proceeding in this fashion, a pattern gradually appears. The general term in the expressions for S_k is conveniently described in a graphical fashion through the use of rooted trees with ordered branches. To each node we associate a scalar valued function, f, and a vector, v, and we refer to the product of the two as the value of the node. The scalar valued function will be given by $H_{0,k}$, $k \geq 2$ or $H_{1,k,m}$ for any k. Here, the second subindex equals the number of branches descending from a particular node, which also equals the number of arguments of f. For the arguments of f we simply substitute the value of the subnodes. If $k = 0$ then the corresponding vector is simply m; otherwise, it is the sum of the vectors at each of the subnodes and m. The tree evaluates to the value of the scalar valued function at the root. A sample tree satisfying these rules is shown below. If we write out the result of evaluating this tree we obtain

$$H_{0,2}(H_{1,0,m_1} m_1, H_{1,0,m_2} m_2), \tag{2.97}$$

which we recognize as the numerator in the first term in Eqs. (2.95-2.96). The denominator in Eq. (2.96), including the factor outside the bracket,

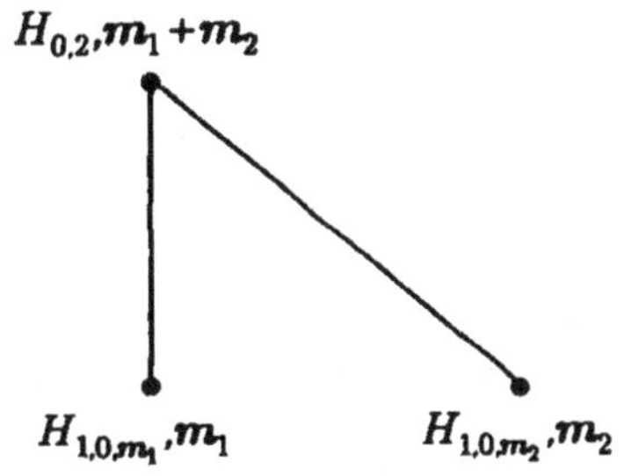

A sample tree corresponding to a term in Ψ_2 and $S_{2,m}$.

is similarly given by a product of factors of the form $H_{0,1}(\boldsymbol{m})$, where $\boldsymbol{m}$ ranges over all the vectors in the tree, whereas in the case of Eq. (2.95) we omit the vector at the root. •

We can further simplify the tree notation by omitting the vector at each node since it is easily generated from the rules outlined. Such a tree is shown below and the rules imply that it corresponds to a term of the form

$$\frac{H_{0,2}(H_{1,1,m_2}(H_{1,0,m_1}\boldsymbol{m}_1)(\boldsymbol{m}_1+\boldsymbol{m}_2), H_{1,0,m_3}\boldsymbol{m}_3)}{H_{0,1}(\boldsymbol{m}_1)H_{0,1}(\boldsymbol{m}_3)H_{0,1}(\boldsymbol{m}_1+\boldsymbol{m}_2)H_{0,1}(\boldsymbol{m}_1+\boldsymbol{m}_2+\boldsymbol{m}_3)}, \tag{2.98}$$

where the last factor in the denominator should be omitted in the case of Ψ. It is

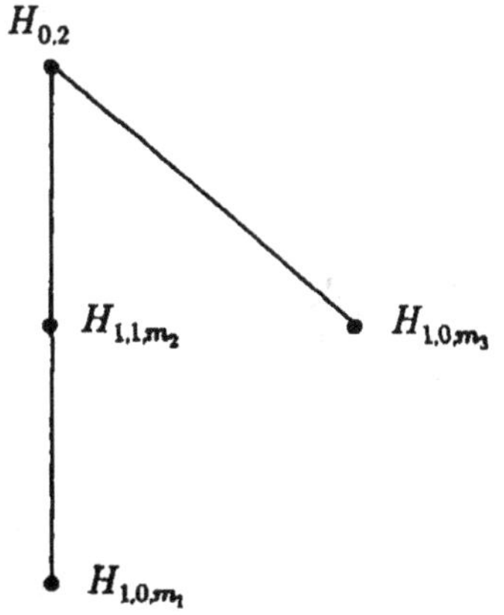

A further simplification in the graphical representation.

easily shown that this term does indeed appear in the expansion for S_3 and the sum is over all vectors $\boldsymbol{m}_i$, $i = 1,2,3$ such that the vector at each node is nonzero. In fact, all the terms in the expressions for S_k and Ψ_k are given by such trees for which the sum of the superscripts at each node equals k. Clearly, the number of trees is a rapidly growing function of k.

We proceed to estimate the terms of the form (2.98) in the Fourier expansions of S_k and Ψ_k. Since H_0 and H_1 are analytic, it follows that

$$|H_{i,j,\boldsymbol{m}}| \leq Cs^{-j}e^{-|\boldsymbol{m}|r}, \tag{2.99}$$

for constants C, $s, r > 0$ independent of i, j, and $\boldsymbol{m}$. Substituting this into the expressions for S_k and Ψ_k, we see that absolute convergence of the Fourier series is guaranteed if

$$|H_{0,1}(\boldsymbol{m})| \geq \frac{1}{\gamma|\boldsymbol{m}|^{\tau}}, \; \forall \boldsymbol{m} \tag{2.100}$$

for some positive constants γ, τ. More specifically, the above condition for $\tau > n-1$, where n equals the number of degrees-of-freedom, is known as the Diophantine condition.

We have argued that the Diophantine condition is sufficient to guarantee absolute convergence of the Fourier series of S_k, albeit not necessarily so for the convergence of the power expansion in ε. Unless there are appropriate cancellations in the Fourier series of S_k so as to eliminate terms with very small denominators, there is no reason to expect the power series to converge. On the contrary, there are terms in S_k that grow much faster than ε^k decays, thus endangering convergence. That such cancellations actually occur in a similar analysis based on the equations of motion was shown through a very careful examination of the different terms in the expansions first performed by Eliasson in the mid-1980s. Such an analysis has yet to be implemented for the Hamilton-Jacobi approach described in this section. Nevertheless, it is expected to carry through, albeit with additional technical complications.

2.4.3 Dynamical consequences

We have described how satisfying the nondegeneracy (2.86) and the Diophantine (2.100) conditions for a given torus implies the existence of a perturbed invariant torus for sufficiently small ε. Moreover, the frequencies of the quasiperiodic motion on the perturbed torus equal those on the unperturbed torus. Naturally, the maximum possible value of ε for which convergence is guaranteed will be different for different unperturbed tori. As $\varepsilon \to 0$ more and more perturbed tori appear and in the limit they have full measure in frequency space, since the Diophantine condition is satisfied on a set of full measure.

For the resonant tori with commensurable unperturbed frequencies, the formal series diverge. It is possible to show that the motion in their vicinity is highly irregular and generally tends to densely fill an open region on the energy-manifold.

Consequently, the motion on the energy manifold is either confined to a perturbed torus, or lies in the stochastic sea away from the tori. In a two-degree-of-freedom system, an energy manifold is three-dimensional and the KAM tori are two-dimensional. It follows that the latter divide the energy manifold into an inside and outside, in such a way as to prohibit motion between the two. The regions between the perturbed tori, known as stochastic layers, are filled with stochastic space-filling motions. For small ε, the stochastic layers are very narrow, since most tori still persist.

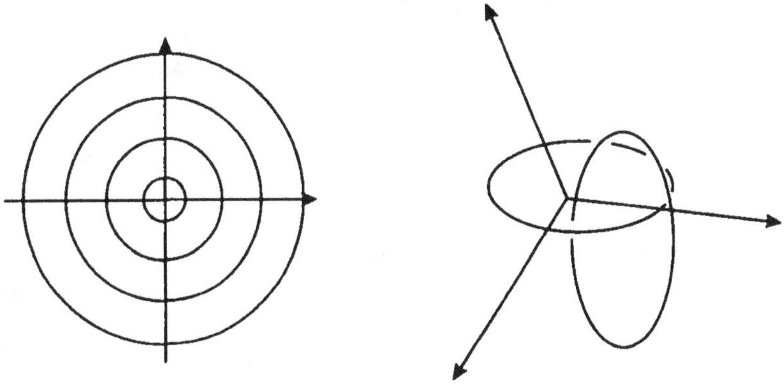

In a two-dimensional space (left panel), one-dimensional tori (circles) confine orbits to remain between these tori. In a higher-dimensional space (right panel), the tori are unable to prevent orbits from reaching all parts of state space.

In higher-degree-of-freedom systems, the n-dimensional tori no longer separate the $2n-1$-dimensional energy manifold. Instead, the stochastic regions near the unperturbed resonant tori extend to form an interconnected web known as the **Arnold web**. Motion along the web, so-called **Arnold diffusion**, is dense on the energy manifold. This effect is believed to persists to $\mathcal{O}(1)$ as $\varepsilon \to 0$, even though the layers become increasingly narrow. It would follow that, even for vanishingly small perturbations, $\mathcal{O}(1)$ displacements from the unperturbed motion can occur given sufficiently long time. One might imagine that different regions of the energy manifold correspond to physical motion of dramatically different character, e.g. corresponding to different excited states in molecules. Through the process of Arnold diffusion, an unexcited molecule could become excited even for very small perturbations. It is, of course, not clear how long this diffusion would take and, consequently, of what importance it is in particular applications.

Questions of this type will reappear in later chapters. We next present the theory behind stochastic, nondeterministic motion in deterministic systems. While the discussion applies to general dynamical systems, the results have important consequences for perturbed Hamiltonian systems. In particular, we will eventually attempt to estimate the Arnold diffusion rate for a particular class of Hamiltonian systems.

2.5 Notes and references

The previous chapter contains some relevant references for the theory described in this chapter. More details on canonical transformations and KAM theory can be found in the work of Lichtenberg & Lieberman[1]. Original proofs of the KAM theorem and many other fascinating topics, of which we will return to in the next chapter, can be found in Moser[2]. Additionally, the review by Chirikov[3] contains a more detailed exposition of Arnold diffusion in theory and experiments.

The modern approach to KAM originating in the work of Eliasson is extensively described in Eliasson[4]. The presentation in this chapter has been chosen to ease the reader into a not always lucid theory. Further generalizations are provided in a series of papers by Gallavotti and collaborators, see Gallavotti[5].

1. A. J. Lichtenberg and M. A. Lieberman, *Regular and Chaotic Dynamics, 2nd ed.* (Springer-Verlag, 1992).
2. J. Moser, *Stable and Random Motions in Dynamical Systems. With Special Emphasis on Celestial Mechanics* (Princeton University Press, 1973).
3. B. V. Chirikov, "A Universal Instability of Many-Dimensional Oscillator Systems," *Physics Reports* **52**(5), (1979), pp. 263-379.
4. L. H. Eliasson, "Absolutely Convergent Series Expansions for Quasiperiodic Motions," *Mathematical Physics Electronic Journal* **3** (1996).
5. G. Gallavotti, "Twistless KAM Tori, Quasi Flat Homoclinic Intersections, and other Cancellations in the Perturbation Series of Certain Completely Integrable Hamiltonian Systems. A Review," *Reviews on Mathematical Physics* **6** (1994), pp. 343-411.

Chapter 3
HOMOCLINIC ORBITS

The satisfactory application of linear, local analysis to engineering problems, and the subsequent disregard for apparently random motion in favor of regular, predictable motion, is highly surprising given the apparent ubiquity of randomness in deterministic dynamics. To those with faith in the workings of nature, regular behavior has an obvious appeal, as it does to those whose future depends on its prediction.

But the pandorian box has been opened. In the previous chapter we saw how highly unpredictable motion was the rule rather than the exception. Among such motions were orbits slowly diffusing through energy manifolds reaching far beyond the reach of the regular motion. In this chapter we discuss an on-the-surface different source of irregular, stochastic behavior, which nevertheless can be shown to be closely related to that in the stochastic layer case. The global dynamical features, whose interaction and consequent impact on state-space dynamics we shall study, are the stable and unstable manifolds of periodic orbits. Section 1 introduces some relevant terminology and investigates the structural stability of intersections of the manifolds, of central importance to the theory to follow.

In Section 2 the **Bernoulli shift** dynamical system is considered. Its range of interesting dynamics serves as a template for properties whose prevalence is the source of so-called chaotic behavior. In fact, the Bernoulli shift is a template in a very concrete way. It can be shown to be in some sense equivalent to a given class of dynamical system, thus warranting its detailed description. Using the theory of exponential splittings, whose properties are described in Section 3, the equivalence is described in detail in Section 4, where the implications of transversal intersections are thoroughly examined.

3.1 General characteristics

Given a hyperbolic saddle point, $\boldsymbol{x}_0$, and its associated stable and unstable manifolds $\mathcal{W}^{s,u}(\boldsymbol{x}_0)$, we call a point common to the manifolds, $\boldsymbol{y} \in \mathcal{W}^s(\boldsymbol{x}_0) \cap \mathcal{W}^u(\boldsymbol{x}_0)$, a **homoclinic point**. Similarly, an orbit $\boldsymbol{y}(t)$ in the intersection of the stable and unstable manifolds of a hyperbolic periodic orbit $\boldsymbol{x}_0(t)$ is called a **homoclinic orbit**. Since points on $\mathcal{W}^s(\boldsymbol{x}_0(t))$ ($\mathcal{W}^u(\boldsymbol{x}_0(t))$) are characterized by their orbits approaching $\boldsymbol{x}_0(t)$ as $t \to \infty$ ($t \to -\infty$), it follows that $\boldsymbol{\phi}(t, \boldsymbol{y}) \to \boldsymbol{x}_0(t)$ as $|t| \to \infty$.

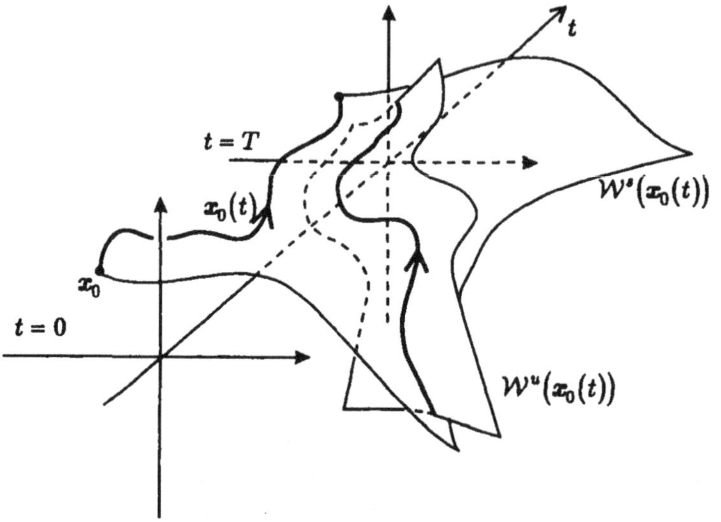

A homoclinic orbit formed by the intersection of the stable and unstable manifolds of a periodic orbit.

The intersection between two manifolds $\mathcal{W}_{1,2}$ is said to be transversal if

$$S = T\mathcal{W}_1 + T\mathcal{W}_2 \tag{3.1}$$

at the point of intersection; i.e., if state space is spanned by their tangent spaces at this point. If the two manifolds are locally given by equations $F_{1,2}(z) = 0$, then the linearization of the system

$$\begin{cases} F_1(z) = 0 \\ F_2(z) = 0 \end{cases} \tag{3.2}$$

has full rank at the point of intersection. Then, for small perturbations of the manifolds, the implicit function theorem implies the existence of a solution in a neighborhood of the point of intersection, such that the intersection is still transversal. In other words, transversal intersections are structurally stable. Their presence ought hence to be of fundamental importance for the behavior of a dynamical system possessing such points. Before we set out to prove this, we introduce a convenient tool whose applications are many.

3.2 The Bernoulli shift

We consider a simple, yet dynamically rich, discrete dynamical system. While its rather dramatic behavior seems nontypical, it is in fact found that this dynamical system is in a sense embedded in a large number of nonlinear dynamical systems.

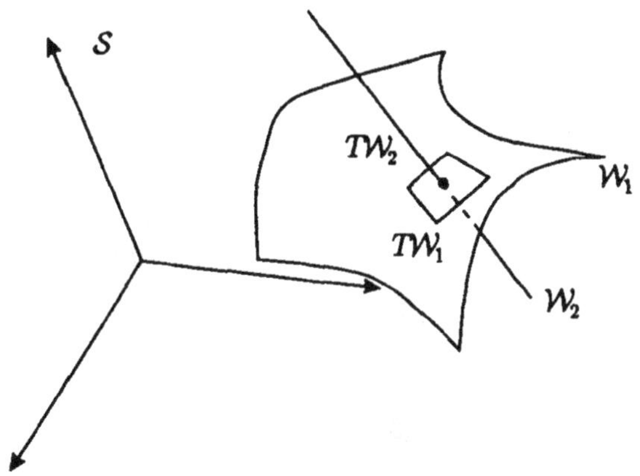

The tangent spaces of two transversally intersecting manifolds span state space.

3.2.1 Topology

Consider an alphabet A of N different symbols. Then, Σ_A denotes the space of bi-infinite sequences

$$\boldsymbol{a} = (\cdots, a_{-3}, a_{-2}, a_{-1}; a_0, a_1, a_2, a_3, \cdots)$$

where $a_k \in A$. A convenient topology on Σ_A is obtained from the metric

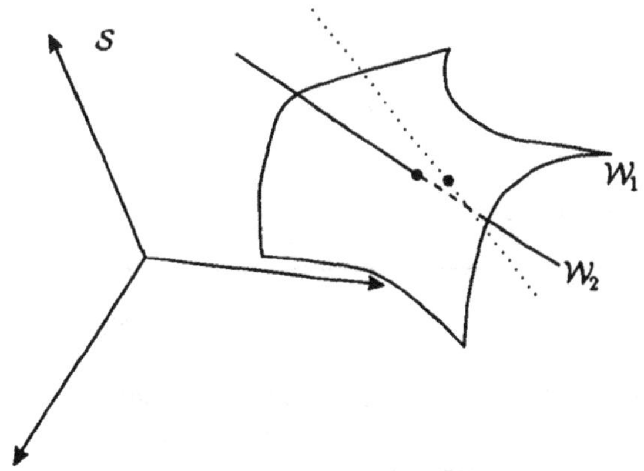

Systems with transversally intersecting manifolds retain this property under small perturbations.

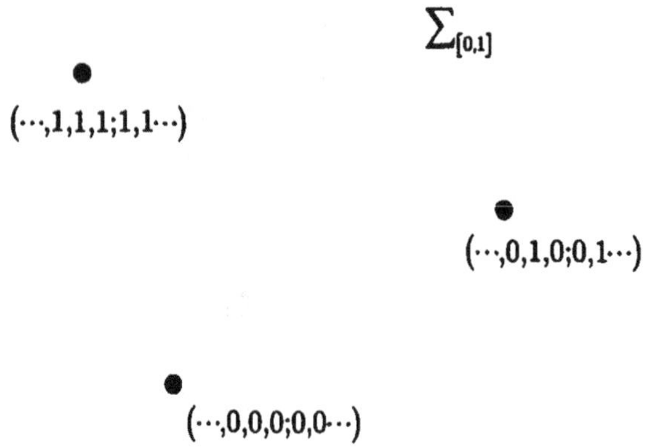

The state space corresponding to the alphabet $A = \{0, 1\}$.

$$d(\boldsymbol{a}, \boldsymbol{a}') = \sum_{n=-\infty}^{\infty} \frac{\delta(a_n, a'_n)}{2^{2|n|+1}} \tag{3.3}$$

where $\delta(x, y) = 0$ when $x = y$ and 1 otherwise. Clearly, the metric emphasizes differences in the components with indices close to zero. If we introduce the family $\{\mathcal{I}_l\}_0^\infty$ of disjoint sets

$$\mathcal{I}_0 = [\frac{1}{2}, \infty) \text{ and } \mathcal{I}_l = [\frac{4^{-l}}{2}, \frac{4^{-l+1}}{3}], \tag{3.4}$$

it follows that

$$a_k = a'_k, |k| \leq l-1 \text{ and } a_k \neq a'_k, k = l \text{ or } -l \Leftrightarrow d(\boldsymbol{a}, \boldsymbol{a}') \in \mathcal{I}_l. \tag{3.5}$$

Distances between points in Σ_A *can lie only within the intervals* $\mathcal{I}_i$.

We show that Σ_A is a Cantor set:

a) Σ_A *is compact.*

Consider an infinite subset $\mathcal{E}$ of Σ_A. Then we can find an infinite subset $\mathcal{E}_0 \subseteq \mathcal{E}$ such that all elements in $\mathcal{E}_0$ have the same value for the zeroth component. In the same way we can obtain an infinite subset $\mathcal{E}_l$ such that all elements in $\mathcal{E}_l$ agree for $|k| \leq l$ and the diameter of $\mathcal{E}_l$ is smaller than $4^{-l+1}/3$. Proceeding in this fashion, we construct an element $\boldsymbol{a} \in \Sigma_A$ such that every neighborhood of $\boldsymbol{a}$ contains an infinite number of elements in $\mathcal{E}$, i.e., $\boldsymbol{a}$ is a limit point of $\mathcal{E}$.

b) Σ_A *is totally disconnected.*

Given two points $\boldsymbol{a}$, $\boldsymbol{a}'$, such that $d(\boldsymbol{a}, \boldsymbol{a}') \in \mathcal{I}_k$, we define the disjoint open sets

$$\mathcal{U} = \{\boldsymbol{b} \mid d(\boldsymbol{a}, \boldsymbol{b}) < \frac{5}{12} 4^{-k}\} \text{ and } \mathcal{V} = \{\boldsymbol{b} \mid d(\boldsymbol{a}, \boldsymbol{b}) > \frac{5}{12} 4^{-k}\}. \tag{3.6}$$

Clearly, $\boldsymbol{a} \in \mathcal{U}$ and $\boldsymbol{a}' \in \mathcal{V}$. Thus, Σ_A is Hausdorff. But it also follows that

$$(\mathcal{U} \cap \Sigma_A) \cup (\mathcal{V} \cap \Sigma_A) = \Sigma_A \tag{3.7}$$

i.e., Σ_A is totally disconnected.

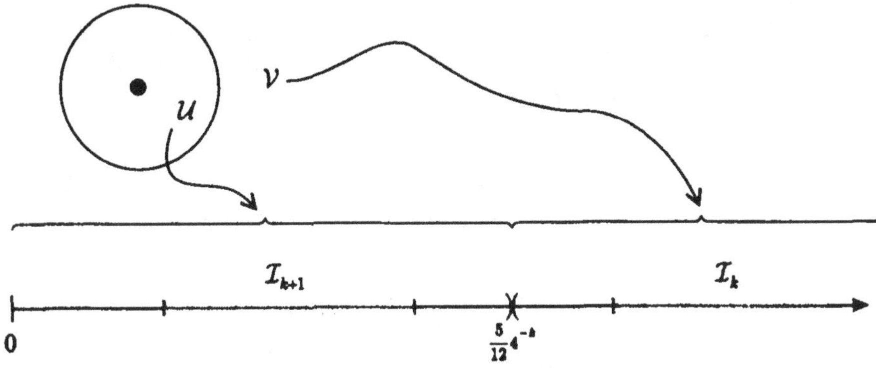

State space can be covered by two disjoint sets $\mathcal{U}$ and $\mathcal{V}$.

c) Σ_A *contains no isolated points.*

Given any point $\boldsymbol{a} \in \Sigma_A$, we can construct a sequence of points $\{\boldsymbol{a}^j\}$ such that $d(\boldsymbol{a}, \boldsymbol{a}^j) \to 0$ as $j \to \infty$. We simply let

$$a_k^j = a_k, \text{ for } |k| \leq j. \tag{3.8}$$

Thus, all points in Σ_A are limit points of Σ_A; i.e., Σ_A contains no isolated points.

With this understanding of the geometry of state space, we proceed to impose a suitable dynamical evolution on Σ_A.

3.2.2 Dynamics

The evolution rule is simply given by the left shift map, σ:

$$(\sigma(a))_k = a_{k+1}. \tag{3.9}$$

We note that the future and past of a particular state is completely contained in a and the dynamical system is thus deterministic.

A glimpse of the complex behavior of this dynamical system is offered by the observation that it possesses periodic orbits of arbitrary period. In particular,

$$\sigma^n(a) = a \Leftrightarrow a_{k+n'} = a_k, \forall k \tag{3.10}$$

where n' is a divisor of n. The number of such points equals the number of ways of filling n spaces with N symbols allowing for repetition, i.e., N^n. We denote by $P(i)$ the number of periodic points with fundamental period i. Thus

$$N^n = \sum_{n'} P(n'), \forall n' \text{ which divide } n. \tag{3.11}$$

Starting with $P(1) = N$ we can solve this equation iteratively for any n. The number of distinct periodic orbits is simply $P(n)/n$, since cyclical permutations of the components correspond to the same orbit.

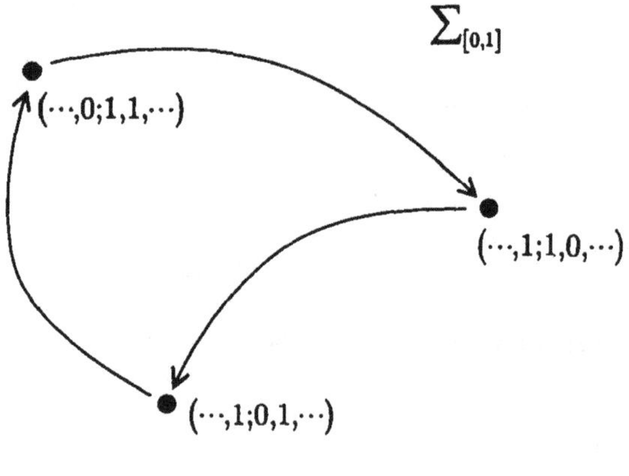

An orbit of period three. Note that the ··· *denote that the pattern is repeated* ad infinitum.

Illustration

The periodic orbits are actually dense in Σ_A. *Given an* $\epsilon > 0$ *and a point* $a \in \Sigma_A$

$$a = (\cdots, a_{-k}, \cdots; a_0, \cdots, a_k, \cdots) \tag{3.12}$$

we simply let $\bar{\boldsymbol{a}}^l$ be a periodic sequence with period $2l+1$, which agrees with $\boldsymbol{a}$ for $|k| \leq l$:

$$\bar{\boldsymbol{a}}^l = (\cdots, \underline{a_{-l}, \cdots; a_0, \cdots, a_l}, \cdots) \tag{3.13}$$

and lies within $4^{-l+1}/3$ from $\boldsymbol{a}$. Clearly, all $\bar{\boldsymbol{a}}^l$ for sufficiently large l lie within ϵ from $\boldsymbol{a}$. •

The inset (outset) of a periodic point $\boldsymbol{a}$ is the set of all eventually periodic sequences in the forward (backward) direction with the same basic building block. One can easily imagine points common to the inset and outset that approach the orbit of $\boldsymbol{a}$ in forward and backward time. There are also nonperiodic orbits. For example,

$$\boldsymbol{a} = (\cdots, a, a, a, b, a, a, b, a, b; a, a, b, a, a, a, b, a, a, a, a, b, \cdots) \tag{3.14}$$

is not periodic under $\boldsymbol{\sigma}$ for any period.

Finally, we show that there is at least one dense orbit in Σ_A under the action of $\boldsymbol{\sigma}$. In particular, we consider all blocks of length 1, length 2, length 3, and so on, and write these one after the other starting at the zeroth position in both directions. It then follows that a forward iterate of this sequence comes arbitrarily close to any point in Σ_A. In fact, one can show that a large majority of orbits in Σ_A are dense. This is essentially the statement that any arbitrary irrational number in $[0, 1)$ contains any arbitrarily prescribed finite block of integers.

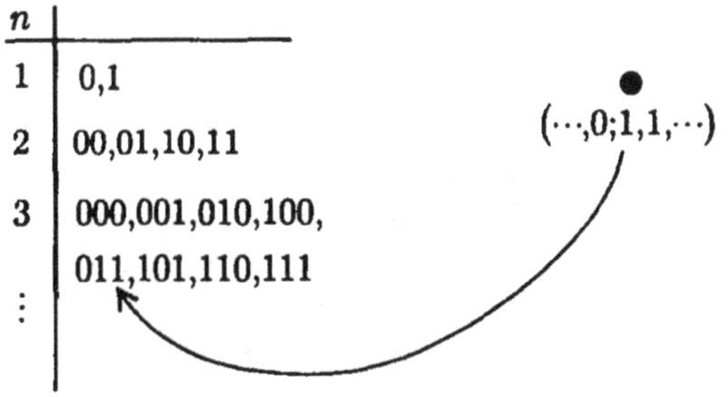

An iterate of the state obtained by concatenating the segments in the table can be made to lie arbitrarily close to any other state.

3.2.3 Implications

The Bernoulli shift would be a mere curiosity were it not for the fact that its behavior is in some sense typical of many nonlinear dynamical systems. If there exists a homeomorphism $\pi : S \to T$ such that

$$\boldsymbol{f}_T = \pi \circ \boldsymbol{f}_S \circ \pi^{-1} \tag{3.15}$$

where $\boldsymbol{f}_S$ is a discrete dynamical system on S, etc., then $\boldsymbol{f}_S$ and $\boldsymbol{f}_T$ are said to be **topologically conjugate**. It follows that

$$\boldsymbol{f}_T^m = \pi \circ \boldsymbol{f}_S^m \circ \pi^{-1} \tag{3.16}$$

for all m; i.e., an orbit of $\boldsymbol{f}_S$ is taken one-to-one and orientation preserving to one of $\boldsymbol{f}_T$. Furthermore, neighborhoods in S are mapped under π onto neighborhoods in T and topological statements are thus easily translated between the two spaces. We note also, that if Image(π) $= \Omega$ is a subset of T, then the we speak of a topological conjugacy *on* Ω.

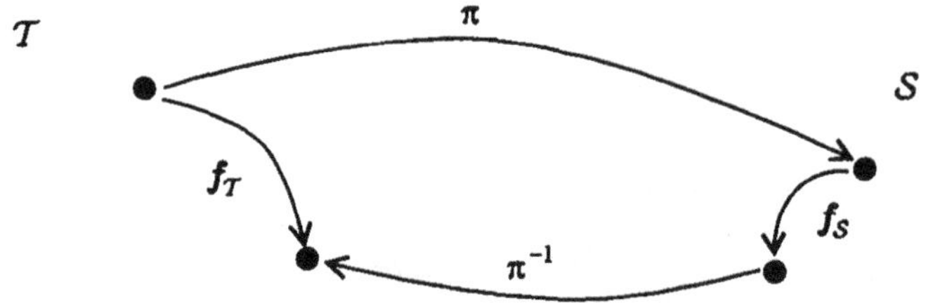

Topological conjugacy allows conclusions to be drawn about one system by refering to its conjugate.

Now consider a periodic continuous dynamical system with its associated period map $\boldsymbol{f}_T$. If we can prove the toplogical conjugacy on a compact set Ω of an iterate of $\boldsymbol{f}_T$ and the shift map $\boldsymbol{\sigma}$ on Σ_A for some suitable A, then all the conclusions for the dynamics of $\boldsymbol{\sigma}$ carry over to those of $\boldsymbol{f}_T$ on Ω. In particular, Ω is an invariant set under $\boldsymbol{f}_T$, which contains unstable periodic points of arbitrary period such that the periodic points are dense. Further, Ω contains uncountably many nonperiodic orbits, as well as orbits that are dense in Ω. It follows that points arbitrarily close to each other may have drastically different evolution. This is known as sensitive dependence on initial conditions. We refer to these characteristics collectively as **deterministic chaos**, and dynamical systems exhibiting them are known as chaotic.

We now proceed to prove the existence of a topological conjugacy between an iterate of the period map and the shift map in the presence of transversal intersections between stable and unstable manifolds of a hyperbolic periodic orbit. This requires the introduction of particular properties of linear dynamical systems to which we now turn.

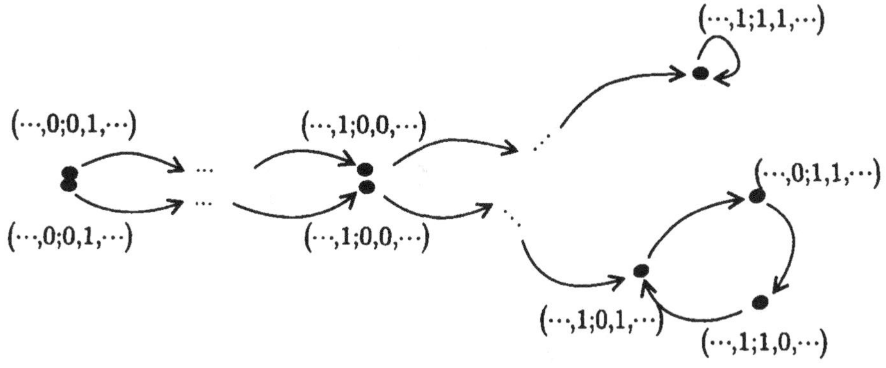

Arbitrarily close to any orbit in state space there is an orbit whose eventual fate lies in the fixed point $(\cdots, 1; 1, 1, \cdots)$.

3.3 Exponential splitting

3.3.1 Definitions and properties

We now generalize the observations in Chapter 1 on autonomous and time-periodic linear systems of hyperbolic type. In particular, we say that a linear dynamical system

$$\dot{\boldsymbol{x}} = \boldsymbol{A}(t)\boldsymbol{x} \tag{3.17}$$

where $\boldsymbol{A}(t)$ is piecewise smooth, admits an **exponential splitting** on $[a, b]$ if there exist two complementary families of subspaces, $\mathcal{E}(s)$, $\mathcal{E}^{\bullet}(s)$, and a corresponding projection operator $\boldsymbol{P}(s)$ such that

$$|\boldsymbol{\phi}(t, s) \circ \boldsymbol{P}(s)| \leq K e^{-\lambda(t-s)}, \text{ for } a \leq s \leq t \leq b \tag{3.18}$$

and

$$|\boldsymbol{\phi}(t, s) \circ (\boldsymbol{I} - \boldsymbol{P}(s))| \leq K e^{-\lambda(s-t)}, \text{ for } a \leq t \leq s \leq b. \tag{3.19}$$

Here, $K \geq 1$ and $\lambda > 0$. In other words, we require that the component of a vector in $\mathcal{E}(s)$ $(\mathcal{E}^{\bullet}(s))$ decays at least exponentially in forward (backward) time. From the discussion in Chapter 1 it follows that the autonomous and time-periodic cases with a hyperbolic fixed point and periodic orbit, respectively, at the origin, admit exponential splittings on $(-\infty, \infty)$.

Given a point $\boldsymbol{y}$ in the stable manifold of a stationary point of a nonlinear autonomous flow, or of the period map of a nonlinear periodic flow, we further argued that the variational equations

$$\dot{\boldsymbol{x}} = \boldsymbol{f}_{\boldsymbol{x}}(\boldsymbol{\phi}(t, \boldsymbol{y}))\boldsymbol{x} \tag{3.20}$$

and

$$\dot{\boldsymbol{x}} = \boldsymbol{f}_{\boldsymbol{x}}(t, \boldsymbol{\phi}(t, \boldsymbol{y}))\boldsymbol{x} \tag{3.21}$$

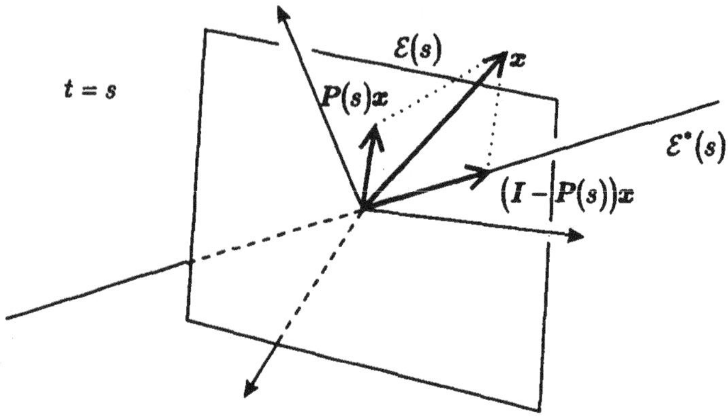

At every instance of time the dynamical system implies a splitting of state space.

respectively, admitted an exponential splitting on $(0, \infty)$. Similarly, if y lay in the unstable manifolds, the variational equations would admit exponential splittings on $(-\infty, 0)$. In the particular case of interest in the present chapter, y actually lies in the intersection between the stable and unstable manifolds. If the intersection at y is transversal; i.e., if

$$\mathcal{S} = T_y\mathcal{W}^s(x_0) + T_y\mathcal{W}^u(x_0), \tag{3.22}$$

it is reasonable to expect that the exponential splittings on $(-\infty, 0)$ and $(0, \infty)$ could be matched at y. This actually can be proven.

Systems that admit exponential splittings satisfy a number of important properties, which we state here without proof.

a) *Robustness*

Assume that (3.17) admits an exponential splitting on (a, b). Then there exists an $\epsilon_0 > 0$ such that the linear system

$$\dot{x} = B(t)x \tag{3.23}$$

also admits an exponential splitting on (a, b) for all $B(t)$ such that

$$\sup_t |A(t) - B(t)| \leq \epsilon < \epsilon_0. \tag{3.24}$$

Moreover, $|P_B(t) - P_A(t)| = \mathcal{O}(\epsilon)$, where P_A is the projection corresponding to the splitting for $A(t)$, etc.

Illustration

In autonomous linear systems with hyperbolic equilibria, all eigenvalues of $\boldsymbol{A}$ are bounded away from the imaginary axis. One can show that the eigenvalues depend continuously on the entries of the matrix. Thus, for sufficiently small (time-independent) perturbations of $\boldsymbol{A}$, the eigenvalues remain bounded away from the imaginary axis. Consequently, the perturbed systems still admit exponential splittings. Similarly, the eigenspaces are only slightly perturbed. It follows that the projection operators are close. •

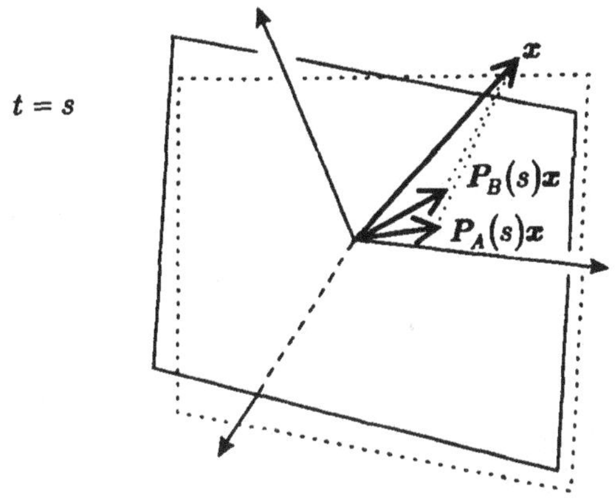

Exponential splittings are structurally stable.

b) *Extension*

If (3.17) admits an exponential splitting on (a, b), then it also admits an exponential splitting on any bounded interval $\mathcal{I} \supseteq (a, b)$. Furthermore, if (a, b) is unbounded to the right (left), then $\mathcal{I}$ needs only to be bounded to the left (right). Denote the projection operator for the extension by $\tilde{\boldsymbol{P}}(t)$. Then, for all $t \in (a, b)$, $\boldsymbol{P}(t)$ and $\tilde{\boldsymbol{P}}(t)$ project onto the same subspace and

$$\begin{aligned} |\tilde{\boldsymbol{P}}(t) - \boldsymbol{P}(t)| &= |\boldsymbol{\phi}(t,a) \circ \tilde{\boldsymbol{P}}(a) \circ (\boldsymbol{I} - \boldsymbol{P}(a)) \circ \boldsymbol{\phi}(a,t)| \\ &\leq |\boldsymbol{\phi}(t,a) \circ \tilde{\boldsymbol{P}}(a)|\,|(\boldsymbol{I} - \boldsymbol{P}(a)) \circ \boldsymbol{\phi}(a,t)| \\ &= |\boldsymbol{\phi}(t,a) \circ \tilde{\boldsymbol{P}}(a)|\,|\boldsymbol{\phi}(a,t) \circ (\boldsymbol{I} - \boldsymbol{P}(t))| \leq K' e^{-\lambda'(t-a)} \end{aligned} \tag{3.25}$$

for some $K' \geq 1$, $\lambda' > 0$.

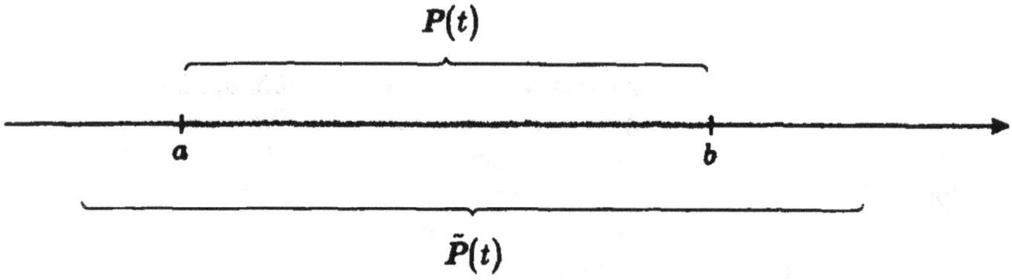

Exponential splittings can be extended to bounded or semibounded intervals.

Illustration

Assume that (3.17) admits an exponential splitting on (a, ∞). One can then show that the linear dynamical system

$$\dot{\boldsymbol{x}} = \boldsymbol{B}(t)\boldsymbol{x}, \tag{3.26}$$

where $\boldsymbol{B}(t) \to \boldsymbol{A}(t)$ as $t \to \infty$, admits an exponential splitting on (a, ∞). In fact, for sufficiently large τ, the robustness property implies that (3.26) admits an exponential splitting on (τ, ∞). Furthermore,

$$|\boldsymbol{P}_{\boldsymbol{B}}(t) - \boldsymbol{P}_{\boldsymbol{A}}(t)| = o(\tau^{-1}) \tag{3.27}$$

for $t > \tau$. Extending the splitting to the original interval (a, ∞), we then find

$$\begin{aligned} |\tilde{\boldsymbol{P}}_{\boldsymbol{B}}(t) - \boldsymbol{P}_{\boldsymbol{A}}(t)| &\le |\tilde{\boldsymbol{P}}_{\boldsymbol{B}}(t) - \boldsymbol{P}_{\boldsymbol{B}}(t)| + |\boldsymbol{P}_{\boldsymbol{B}}(t) - \boldsymbol{P}_{\boldsymbol{A}}(t)| \\ &\le K' e^{-\lambda'(t-\tau)} + o(\tau^{-1}) \end{aligned} \tag{3.28}$$

for $t > \tau$. Restricting attention to $t > 2\tau$, the right-hand side becomes bounded by $K' e^{-\lambda' \tau} + o(\tau^{-1})$ and hence

$$|\tilde{\boldsymbol{P}}_{\boldsymbol{B}}(t) - \boldsymbol{P}_{\boldsymbol{A}}(t)| \to 0, t \to \infty. \tag{3.29}$$

•

c) *Concatenation*

Consider a partition of the real line of intervals $\Delta_k = [t_{k-1}, t_k]$ such that

$$\inf_k |\Delta_k| = \frac{\tau}{\lambda} \ln K \tag{3.30}$$

and a corresponding sequence $\{A_k(t)\}$, each of which admits an exponential splitting on Δ_k. We refer to the system

$$\dot{x} = A(t)x, \tag{3.31}$$

where $A(t) = A_k(t)$, $t \in \Delta_k$, as the concatenation of $\{A_k(t)\}$. It follows that

$$|\phi(t_k, t_{k-1}) \circ P_k(t_{k-1})| \le Ke^{-\lambda|\Delta_k|} = K^{1-r}, \tag{3.32}$$

which is less than 1 if $r > 1$. It is now possible to show that, provided the subspace decompositions match up at each end point and that $r > 1$, the concatenated system admits an exponential splitting on $(-\infty, \infty)$.

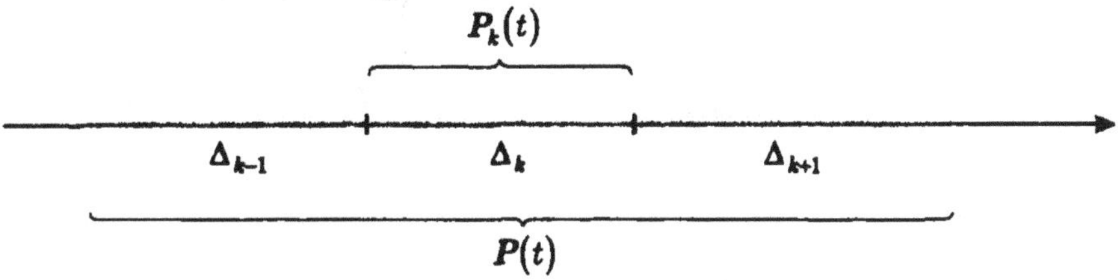

Piecewise exponential splittings can be merged to a single exponential splitting.

Illustration

It is not necessary to match the subspace decompositions exactly at each end point. Assuming that the $A_k(t)$ are uniformly bounded, there exists an $\epsilon_0 > 0$ such that the concatenation admits an exponential splitting on $(-\infty, \infty)$ if

$$|P_k(t_{k-1}) - P_{k-1}(t_{k-1})| \le \epsilon < \epsilon_0. \tag{3.33}$$

The proof essentially consists of reducing the problem to the previous case by constructing a sequence $\{B_k(t)\}$, admitting exponential splittings on Δ_k, with matching subspace decompositions. In fact,

$$\sup_t |B(t) - A(t)| = o(\epsilon). \tag{3.34}$$

Thus, choosing small enough ϵ_0 allows the use of robustness to show that the original concatenation admits an exponential splitting on $(-\infty, \infty)$.

•

d) *Nonlinearity*

Consider adding a nonlinear time-dependent perturbation to the linear vector field in Eq. (3.17):

$$\dot{\boldsymbol{x}} = \boldsymbol{A}(t)\boldsymbol{x} + \boldsymbol{g}(t, \boldsymbol{x}). \tag{3.35}$$

If, for a given $\epsilon > 0$, there exists a $\delta > 0$ such that

$$|\boldsymbol{g}(t, \boldsymbol{0})| \leq \delta, \text{ and } |\boldsymbol{g}_{\boldsymbol{x}}(t, \boldsymbol{x})| = o(\boldsymbol{x}), \tag{3.36}$$

then there exists a unique solution $\boldsymbol{x}(t)$ to Eq. (3.35) such that

$$\sup_{-\infty \leq t \leq \infty} |\boldsymbol{x}(t)| = \min\{2\epsilon, 4\frac{K}{\lambda}\delta\}. \tag{3.37}$$

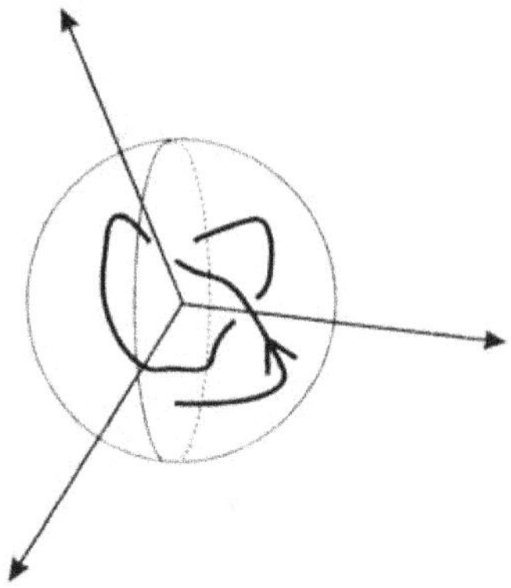

Under nonlinear perturbations of an exponential splitting, there is a unique orbit which remains within a neighborhood of the origin.

Illustration

Let $\boldsymbol{z}(t)$ and $\boldsymbol{y}(t)$ be two solutions to the differential equation

$$\dot{\boldsymbol{x}} = \boldsymbol{f}(t, \boldsymbol{x}) \tag{3.38}$$

such that the variational equation

$$\dot{\boldsymbol{x}} = \boldsymbol{f}_{\boldsymbol{x}}(t, \boldsymbol{z}(t))\boldsymbol{x} \tag{3.39}$$

admits an exponential splitting on $(-\infty, \infty)$. Let $\boldsymbol{x}(t) = \boldsymbol{y}(t) - \boldsymbol{z}(t)$. Then, from Eq. (3.38) we find that $\boldsymbol{x}(t)$ satisfies

$$\dot{\boldsymbol{x}} = \boldsymbol{f}_{\boldsymbol{x}}(t, \boldsymbol{z}(t))\boldsymbol{x} + \boldsymbol{g}(t, \boldsymbol{x}) \tag{3.40}$$

where

$$\boldsymbol{g}(t, \boldsymbol{x}) = \boldsymbol{f}(t, \boldsymbol{z}(t) + \boldsymbol{x}) - \boldsymbol{f}(t, \boldsymbol{z}(t)) - \boldsymbol{f}_{\boldsymbol{x}}(t, \boldsymbol{z}(t))\boldsymbol{x}. \tag{3.41}$$

We immediately see that $\boldsymbol{g}(t, \boldsymbol{0}) = \boldsymbol{0}$ and

$$|\boldsymbol{g}_{\boldsymbol{x}}(t, \boldsymbol{x})| = |\boldsymbol{f}_{\boldsymbol{x}}(t, \boldsymbol{z}(t) + \boldsymbol{x}) - \boldsymbol{f}_{\boldsymbol{x}}(t, \boldsymbol{z}(t))|, \tag{3.42}$$

which is $o(\boldsymbol{x})$ if $\boldsymbol{f}_{\boldsymbol{x}}$ is uniformly continuous in $\boldsymbol{x}$ with respect to both t and $\boldsymbol{x}$. The above result then implies that there is a unique solution that stays near $\boldsymbol{0}$ for all times. But, $\boldsymbol{x}(t) = \boldsymbol{0}$ is in fact a solution, and hence the only solution in a neighborhood of the origin. Thus, there are no other solutions $\boldsymbol{y}(t)$ of (3.38) that stay close to $\boldsymbol{z}(t)$ for all times. The applications to autonomous and time-periodic systems should be obvious.

•

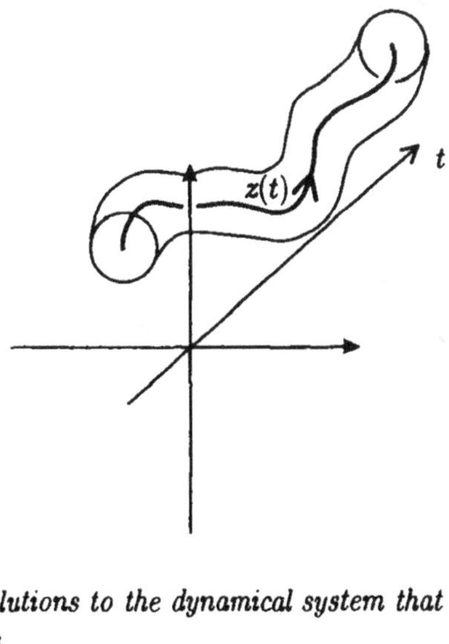

There exist no other solutions to the dynamical system that stay within the cylinder about $\boldsymbol{z}(t)$ for all times.

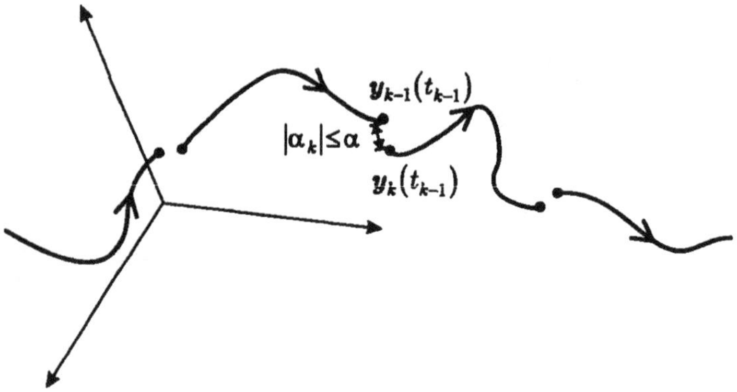

An α-pseudo orbit consists of collections of orbit segments.

3.3.2 Shadowing orbits

Consider a partition of the real line into intervals $\Delta_k = [t_{k-1}, t_k]$. We refer to a corresponding sequence $\{\boldsymbol{y}_k(t)\}$ of solutions to the equation

$$\dot{\boldsymbol{x}} = \boldsymbol{f}(t, \boldsymbol{x}) \tag{3.43}$$

such that $|\boldsymbol{y}_k(t_{k-1}) - \boldsymbol{y}_{k-1}(t_{k-1})| = |\alpha_k| < \alpha$ as an α-pseudo-orbit for $\boldsymbol{f}$. The orbit $\boldsymbol{y}(t)$ is said to β-shadow $\{\boldsymbol{y}_k(t)\}$ if

$$|\boldsymbol{y}(t) - \boldsymbol{y}_k(t)| < \beta,\ t \in \Delta_k, \forall k. \tag{3.44}$$

In other words, the solution $\boldsymbol{y}(t)$ is approximated by the sequence $\{\boldsymbol{y}_k(t)\}$ within β

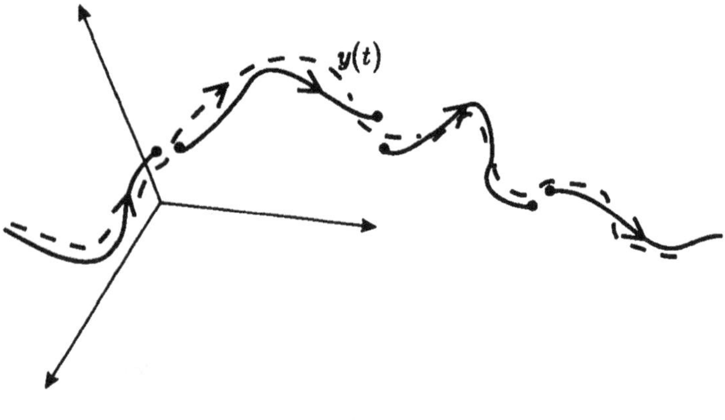

A β-shadowing orbit interpolates an α-pseudo orbit.

for all time. We aim to show the existence of such a shadowing orbit under certain conditions on $\boldsymbol{f}$ and $\{\boldsymbol{y}_k(t)\}$.

Assume that $\boldsymbol{A}_k(t) = \boldsymbol{f}_{\boldsymbol{x}}(t, \boldsymbol{y}_k(t))$ is uniformly bounded in k, where $\boldsymbol{f}_{\boldsymbol{x}}$ is uniformly continuous in $\boldsymbol{x}$. Further, assume that the flow corresponding to $\boldsymbol{A}_k$ admits an exponential splitting on Δ_k such that

$$|\boldsymbol{P}_k(t_{k-1}) - \boldsymbol{P}_{k-1}(t_{k-1})| \leq \epsilon < \epsilon_0, \tag{3.45}$$

where ϵ_0 has been chosen as in Eq. (3.33). In the notation of that section it then follows that the system corresponding to $\boldsymbol{A}(t)$ admits an exponential splitting on $(-\infty, \infty)$. By robustness,

$$\dot{\boldsymbol{x}} = \boldsymbol{f}_{\boldsymbol{x}}(t, \boldsymbol{z}(t))\boldsymbol{x}, \tag{3.46}$$

where $\boldsymbol{z}(t)$ is a continuous function for which $|\boldsymbol{z}(t) - \boldsymbol{y}_k(t)| < \alpha$, also admits an exponential splitting on $(-\infty, \infty)$ provided α is sufficiently small. Such a $\boldsymbol{z}(t)$ is provided by

$$\boldsymbol{z}(t) = \boldsymbol{y}_k(t) + \frac{\alpha_k}{2}\frac{t - t_{k-1}}{t_k - t_{k-1}} + \frac{\alpha_{k-1}}{2}\frac{t - t_k}{t_k - t_{k-1}}, \; t \in \Delta_k, \forall k. \tag{3.47}$$

This function has a continuous derivative on the interior of Δ_k where, in fact,

$$|\dot{\boldsymbol{z}}(t) - \dot{\boldsymbol{y}}_k(t)| \leq \alpha(\inf_k |\Delta_k|)^{-1}. \tag{3.48}$$

We assume that the interval size is such that this quantity is less than α.

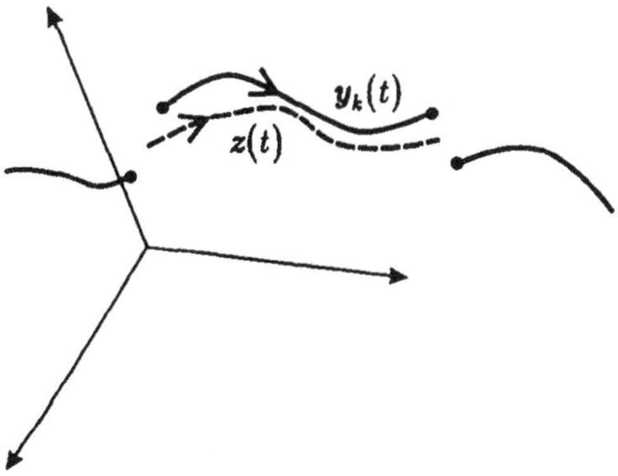

The first step in interpolating the α-pseudo orbit.

Now, assume that $\boldsymbol{y}(t)$ is a solution to Eq. (3.43) and define $\boldsymbol{x}(t) = \boldsymbol{y}(t) - \boldsymbol{z}(t)$. Then $\boldsymbol{x}(t)$ satisfies

$$\dot{\boldsymbol{x}} = \boldsymbol{f}_{\boldsymbol{x}}(t, \boldsymbol{z}(t))\boldsymbol{x} + \boldsymbol{g}(t, \boldsymbol{x}) \tag{3.49}$$

where

$$g(t,x) = -\dot{z}(t) + f(t, z(t) + x) - f_x(t, z(t))x. \tag{3.50}$$

Now,

$$|g(t,0)| \le |f(t,z(t)) - f(t,y_k(t))| + |\dot{y}_k(t) - \dot{z}(t)| \le C\alpha, \tag{3.51}$$

for all t in the interior of Δ_k. Furthermore,

$$|g_x(t,x)| = |f_x(t,z(t)+x) - f_x(t,z(t))| \tag{3.52}$$

which is $o(x)$ since f_x is uniformly continuous.

The results of the previous section now imply the existence of a unique solution $x(t)$ to Eq. (3.49) such that

$$\sup_t |x(t)| = \min\{2\beta, \frac{4K}{\lambda} \text{ ess sup}_t |g(t,0)|\}. \tag{3.53}$$

Thus, $y(t) = z(t) + x(t)$ is a solution to (3.43) and we can chose α such that

$$|y(t) - y_k(t)| \le |x(t)| + |z(t) - y_k(t)| \le \beta \tag{3.54}$$

i.e., $y(t)$ β-shadows $\{y_k(t)\}$. It is, in fact, unique, since for any other β-shadowing orbit, $\tilde{y}(t) = z(t) + \tilde{x}(t)$, we have that

$$|\tilde{x}(t)| \le |\tilde{y}(t) - y_k(t)| + |y_k(t) - z(t)| \le 2\beta \tag{3.55}$$

for sufficiently small α. But, then $x(t) = \tilde{x}(t)$, which proves uniqueness.

3.4 Transversal intersections

It is now time to apply the results of the previous sections to the case of transversal intersections of the stable and unstable manifolds of a hyperbolic saddle point. In particular, we consider the dynamical system

$$\dot{x} = f(t,x) \tag{3.56}$$

where f is periodic with period T, and moreover has a hyperbolic periodic orbit, $x_0(t)$ of the same period. Then, the variational equation

$$\dot{x} = f_x(t, x_0(t))x \tag{3.57}$$

admits an exponential splitting on $(-\infty, \infty)$.

Furthermore, we assume the existence of an orbit $y(t)$ such that $\phi(t,y) \to x_0(t)$ as $|t| \to \infty$, i.e., $y(t)$ lies in the intersection between the stable and unstable manifolds of $x_0(t)$. If the intersection is transversal, then the variational equation

$$\dot{x} = f_x(t, y(t))x \tag{3.58}$$

also admits an exponential splitting on $(-\infty,\infty)$. Clearly, the same holds true for the equation

$$\dot{\boldsymbol{x}} = \boldsymbol{f}_x(t, \boldsymbol{y}_k(t))\boldsymbol{x} \tag{3.59}$$

where $\boldsymbol{y}_k(t) = \boldsymbol{y}(t + a_k T)$, for some $\boldsymbol{a} \in \Sigma_A$, $\forall k$. Also, $|\boldsymbol{y}_k(t) - \boldsymbol{x}_0(t)| \to 0$ uniformly in k as $|t| \to \infty$. Finally, if $\boldsymbol{P}(t)$ denotes the projection corresponding to the splitting of (3.57) and $\boldsymbol{P}_k(t)$ that of (3.59), then $|\boldsymbol{P}_k(t) - \boldsymbol{P}(t)| \to 0$ uniformly in k as $|t| \to \infty$.

We now introduce the partition Δ_k such that $t_k = 2kmT$, where m is some integer. The corresponding sequence $\{\tilde{\boldsymbol{y}}_k\}$ such that

$$\tilde{\boldsymbol{y}}_k(t) = \boldsymbol{y}_k(t - (2k-1)mT) \tag{3.60}$$

and its corresponding sequence of projections $\tilde{\boldsymbol{P}}_k(t) = \boldsymbol{P}_k(t - (2k-1)mT)$ then satisfy

$$|\tilde{\boldsymbol{y}}_k(t_{k-1}) - \tilde{\boldsymbol{y}}_{k-1}(t_{k-1})| \le |\boldsymbol{y}_k(-mT) - \boldsymbol{x}_0(-mT)| + |\boldsymbol{x}_0(mT) - \boldsymbol{y}_{k-1}(mT)| \tag{3.61}$$

and

$$|\tilde{\boldsymbol{P}}_k(t_{k-1}) - \tilde{\boldsymbol{P}}_{k-1}(t_{k-1})| \le |\boldsymbol{P}_k(-mT) - \boldsymbol{P}(-mT)| + |\boldsymbol{P}(mT) - \boldsymbol{P}_{k-1}(mT)|. \tag{3.62}$$

Assuming that $\boldsymbol{f}_x$ is uniformly bounded, and choosing m sufficiently large, we can satisfy the conditions of the previous section and hence find a unique solution $\boldsymbol{y}_{\boldsymbol{a}}(t)$ to (3.56) such that

$$\sup_{t \in \Delta_k} |\boldsymbol{y}_{\boldsymbol{a}}(t) - \tilde{\boldsymbol{y}}_k(t)| \le \beta \tag{3.63}$$

for all k. In other words,

$$\sup_{t \in (-mT, mT)} |\boldsymbol{y}_{\boldsymbol{a}}(t + (2k-1)mT) - \boldsymbol{y}(t + a_k T)| \le \beta \tag{3.64}$$

for all k.

We have proven the existence of a β-shadowing orbit to the α-pseudo-orbit $\boldsymbol{y}(t + a_k T)$ for any $\boldsymbol{a} \in \Sigma_A$. For example, the point

$$\boldsymbol{a} = (\ldots, 0, 1, 0, 1, 0, 1, 0, 1, \ldots) \tag{3.65}$$

in $\Sigma_{[0,1]}$ yields the sequence $(\ldots, \boldsymbol{y}(t), \boldsymbol{y}(t+T), \ldots,)$ for $t \in (-mT, mT)$. The above results thus imply the existence of a solution to (3.56), which jumps periodically back and forth between the two different elements in this sequence. Similarly, a nonperiodic point under the shift map corresponds to a nonperiodic jumping.

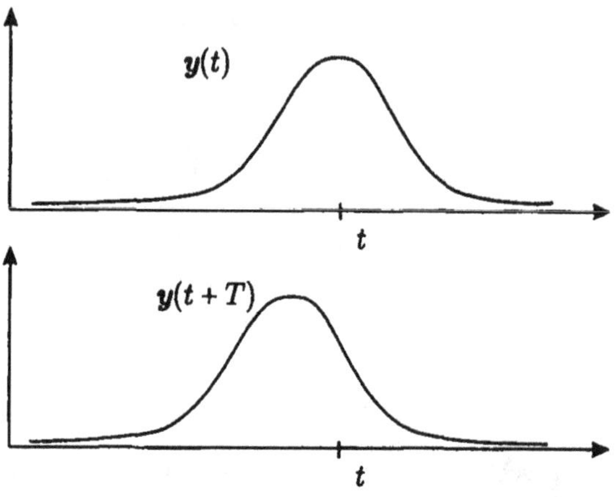

With the alphabet $A = \{0,1\}$ the solutions alternate in arbitrary fashion between shadowing the two segments $\boldsymbol{y}(t)$ and $\boldsymbol{y}(t+T)$.

Illustration

The above discussion also applies to the case where the stable and unstable manifolds of several hyperbolic periodic orbits intersect transversely to form a heteroclinic cycle. One then obtains a β-shadowing orbit, which jumps between segments on each of the branches of the cycle. This provides a means for moving from the vicinity of one hyperbolic orbit to that of another, possibly very far from the original. We refer to such a cycle as a **transition chain** *and the shadowing solution as a* **transition orbit**. *Clearly, the minimum time required for a transition orbit is determined by the integer m, estimates of which allow one to make more general predictions about the distribution of transition times. We will return to issues of this type in later chapters, where transition orbits will have fundamental physical implications.* •

We finally show the existence of a homeomorphism, $\pi : \Sigma_a \rightarrow \Omega$, where $\Omega \in \mathcal{S}$ is compact, such that $\pi \circ \boldsymbol{\sigma} \circ \pi^{-1}$ equals $\boldsymbol{f}_T^{2m}$, where $\boldsymbol{f}_T$ is the period map of (3.56). Define $\pi(\boldsymbol{a}) = \boldsymbol{y}_{\boldsymbol{a}}(0)$. Clearly, $\boldsymbol{y}_{\boldsymbol{a}}(t + 2mT) = \boldsymbol{y}_{\sigma(\boldsymbol{a})}(t)$, since

$$|\boldsymbol{y}_{\boldsymbol{a}}(t + 2mT + (2k-1)mT) - \boldsymbol{y}(t + a_{k+1}T)| \leq \beta \tag{3.66}$$

and $\boldsymbol{y}_{\sigma(\boldsymbol{a})}$ is unique. Thus,

$$\boldsymbol{f}_T^{2m}(\pi(\boldsymbol{a})) = \boldsymbol{y}_{\boldsymbol{a}}(2mT) = \boldsymbol{y}_{\sigma(\boldsymbol{a})}(0) = \pi(\boldsymbol{\sigma}(\boldsymbol{a})). \tag{3.67}$$

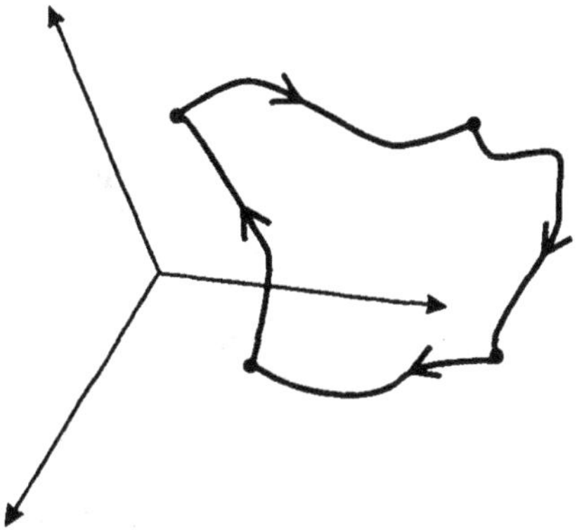

A heteroclinic cycle formed by transversally intersecting manifolds forms a transition chain.

Since Σ_A is compact and Hausdorff, it suffices to show that π is bijective and continuous.

a) *Bijectivity*

Consider two points $\boldsymbol{a} \neq \boldsymbol{a}'$. Then $a_k \neq a'_k$ for some k and

$$|\boldsymbol{y}_{\boldsymbol{a}}(-a_kT + (2k-1)mT) - \boldsymbol{y}_{\boldsymbol{a}'}(-a_kT + (2k-1)mT)|$$

$$\geq |\boldsymbol{y}(0) - \boldsymbol{y}((a'_k - a_k)T)| - |\boldsymbol{y}_{\boldsymbol{a}}(-a_kT + (2k-1)mT) - \boldsymbol{y}(0)|$$
$$-|\boldsymbol{y}_{\boldsymbol{a}'}(-a_kT + (2k-1)mT) - \boldsymbol{y}((a'_k - a_k)T)|. \qquad (3.68)$$

We can choose β sufficiently small, with a corresponding increase in m, so that this expression is greater than β for all t. It then follows that $\boldsymbol{y}_{\boldsymbol{a}}(0) \neq \boldsymbol{y}_{\boldsymbol{a}'}(0)$, i.e., π is bijective.

b) *Continuity*

Consider a sequence $\{\boldsymbol{a}^j\}$ such that $\boldsymbol{a}^j \to \boldsymbol{a}$. Then, we can choose a subsequence (also denoted $\boldsymbol{a}^j$) such that $\{\pi(\boldsymbol{a}^j)\}$ converges to a point $\boldsymbol{x}(0)$, since

$$|\pi(\boldsymbol{a}^j)| \leq |\pi(\boldsymbol{a}^j) - \boldsymbol{y}(a^j_kT - (2k-1)mT)|$$
$$+|\boldsymbol{y}(a^j_kT - (2k-1)mT)| \leq \beta + \sup_t |\boldsymbol{y}(t)| < \infty. \qquad (3.69)$$

Continuous dependence on initial conditions for ordinary differential equations implies that $\boldsymbol{y}_{\boldsymbol{a}^j}(t) \to \boldsymbol{x}(t)$ uniformly on compact intervals in t. For a given k and all $t \in [-mT, mT]$

$$|\boldsymbol{x}(t + (2k-1)mT) - \boldsymbol{y}(t + a_kT)|$$

$$\leq |x(t+(2k-1)mT) - y_{a^j}(t+(2k-1)mT)|$$
$$+|y_{a^j}(t+(2k-1)mT) - y(t+a_kT)|, \qquad (3.70)$$

which clearly goes to zero as $j \to \infty$. But uniqueness of the shadowing orbit then implies that $x(t) = y_a(t)$, i.e., π is continuous.

We have thus shown the topological conjugacy between an iterate of the period map and the Bernoulli shift on Σ_A. In other words, dynamical systems possessing periodic orbits, whose stable and unstable manifolds intersect transversely, exhibit chaotic behavior. For example, their dynamics are highly sensitive to choice of initial conditions. Clearly, the ability to make long-term predictions is drastically reduced. It appears that the detection of transversal intersections would be a vital step in the understanding of global dynamics of nonlinear systems. It is to this task that the next chapter is devoted.

3.5 Liapunov exponents

We have seen how chaotic dynamics are related to rapid divergence of future orbits for near-lying initial points. This is the sensitive dependence on initial conditions that was referred to in Section 2.3. For regular motion, on the other hand, orbits are expected to separate at a more modest pace. This is the justification for the use of numerical simulations for prediction, where one can at best expect a very approximate knowledge of the initial state of the system. In chaotic, stochastic systems, such predictions are inherently doomed to fail. It is, however, possible to make statistical predictions about temporal and spatial averages, the subject of **ergodic theory**.

It appears that a convenient diagnostic means for detecting stochastic behavior is the rate at which orbits separate. We consider two initial points x_0 and $x_0 + \mathrm{d}x$ and their separation at time t:

$$\delta x_0(t) = |\phi(t, x_0 + \mathrm{d}x) - \phi(t, x_0)|. \qquad (3.71)$$

Illustration

Consider a linear autonomous system

$$\dot{x} = Ax \qquad (3.72)$$

with a hyperbolic equilibrium at the origin, $x_0 = \mathbf{0}$. We recall the general solution

$$x = \sum_{k=1}^{m}(a_0^k + a_1^k t + \ldots + a_{\mu_k - 1}^k t^{\mu_k - 1})e^{\lambda_k t} \qquad (3.73)$$

where λ_k are eigenvalues of A with multiplicity μ_k. To each eigenvalue

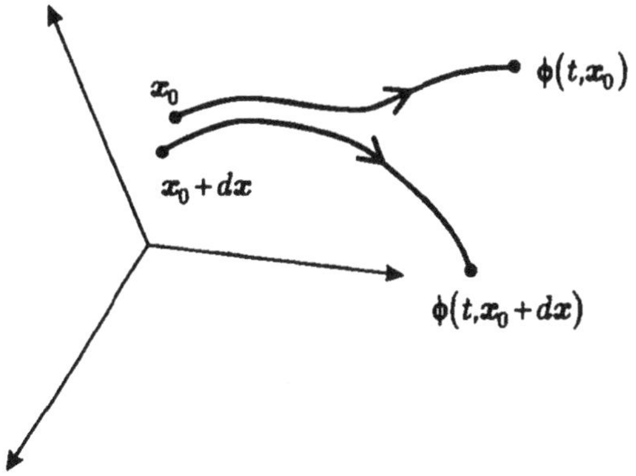

Orbital separation is a measure of the sensitivity to initial conditions.

λ_k there exists an invariant subspace $\mathcal{E}_k$ obtained from the generalized eigenvectors corresponding to λ_k. Thus, for $dx \in \mathcal{E}_k$:

$$\delta x_0(t) = |\phi(t, dx) - \phi(t, 0)| = |a_0^k + a_1^k t + \ldots + a_{\mu_k}^k t^{\mu_k}| e^{\Re(\lambda_k)t} \quad (3.74)$$

and hence

$$\lim_{t\to\infty} \frac{1}{t} \ln \delta x_0(t) = \Re(\lambda_k) \quad (3.75)$$

Clearly, for most initial conditions, dx, the left-hand side will equal the real part of the largest eigenvalue. Subtracting the component in the subspace corresponding to this eigenvalue, we similarly obtain the next largest eigenvalue, and so on. •

Guided by the illustration, we define the rate of separation

$$\sigma(x_0) = \lim_{t\to\infty} \frac{1}{t} \ln \delta x_0(t) \quad (3.76)$$

as $dx \to \mathbf{0}$. It should be noted that σ in general depends on the way in which dx approaches $\mathbf{0}$. Nevertheless, it is possible to show that $\sigma \in \mathcal{N}$, where $\mathcal{N}$ contains at most n different elements. These are known as the **Liapunov exponents** of x_0. Just as in the linear case, along most directions for dx, σ equals the largest member of $\mathcal{N}$. Moreover, there is a hierarchy of subspaces $\mathcal{E}_k \supset \mathcal{E}_{k+1}$ such that for $dx \in \mathcal{E}_k$, σ equals the k-th Liapunov exponent, where we imagine the exponents in descending order.

Since stochastic behavior is characterized by an exponential rate of separation between orbits, we expect the largest Liapunov exponent of an orbit $\phi(t, x_0)$ in a stochastic region to be positive. Similarly, the absence of positive Liapunov exponents would indicate regular motion in the vicinity of the orbit of x_0.

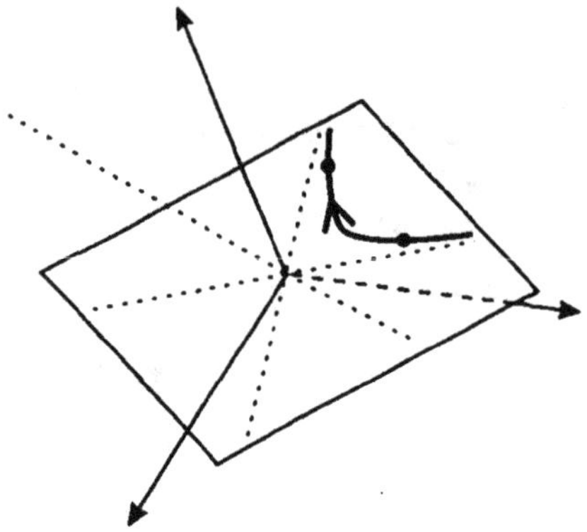

In a linear system, the separation vector grows at an exponential rate determined by the largest eigenvalue.

Illustration

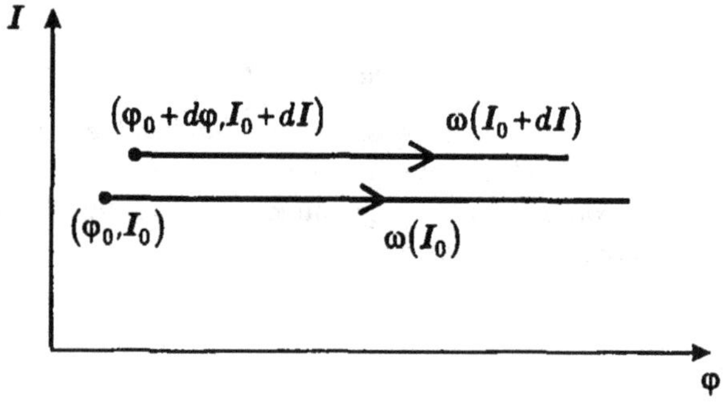

In an integrable Hamiltonian system, the separation vector grows at most linearly.

For an integrable Hamiltonian system in action-angle coordinates we have

$$I(t) = I_0,\ \varphi(t) = \varphi_0 + \omega(I_0)t. \tag{3.77}$$

Consequently, for initial conditions $I_0 + \mathrm{d}I$, $\varphi_0 + \mathrm{d}\varphi$ we find

$$I(t) = I_0 + \mathrm{d}I,\ \varphi(t) = \varphi_0 + \mathrm{d}\varphi + \omega(I_0 + \mathrm{d}I)t, \tag{3.78}$$

which implies

$$\delta(\varphi_0, I_0)(t) = |(\mathrm{d}\varphi + [\omega(I_0 + \mathrm{d}I) - \omega(I_0)]\, t, \mathrm{d}I)|. \tag{3.79}$$

Thus, the separation grows at most linearly in t for large t. Consequently, all Liapunov exponents vanish in the integrable case. That this is not so for many orbits in the nonintegrable case is a property that allows us to numerically investigate integrability. •

The definition of σ requires numerical integrations over large times during which any expansion is likely to cause numerical overflow. This practical difficulty is overcome by repeatedly following the separation for short intervals of time Δt, and normalizing $d\boldsymbol{x}$ at the end of each such interval. Since we are interested in letting $d\boldsymbol{x}$ approach zero, it suffices to consider the linearized flow around the orbit of $\boldsymbol{x}_0$. Finally, one can show that, for sufficiently short Δt,

$$\sigma_{max}(\boldsymbol{x}_0) = \lim_{n\to\infty} \frac{1}{n\Delta t} \sum_{k=1}^{n} \ln \delta_k \boldsymbol{x}_0, \tag{3.80}$$

where $\delta_k \boldsymbol{x}_0$ is the separation at the end of the k-th interval and σ_{max} the largest Liapunov exponent. We will apply this formula in subsequent chapters to further illustrate our analytical results.

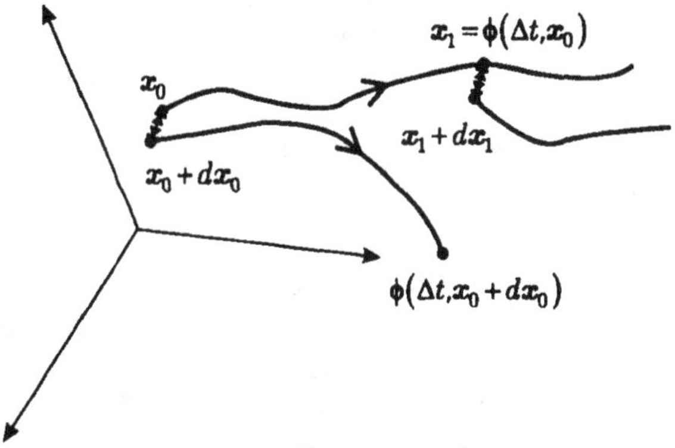

We approximate the Liapunov exponent through an ensemble average obtained by short-time integrations along the original orbit.

3.6 Notes and references

In this chapter the consequences of transversal intersections between stable and unstable manifolds of a periodic orbit were studied. This was achieved by introducing the idea of a Bernoulli shift on infinite symbol sequences and showing the equivalence of the shift with an iterate of the period map using the idea of exponential splittings. Also, a common and useful diagnostic tool for the detection of chaotic behavior was presented in the form of Liapunov exponents.

The shift map and its properties have been known to investigators since the turn of the century. A particularly important implementation of the shift map in geometric form was achieved by Smale[1] in the 1960s. This goes under the name **the Smale horseshoe**, and is perhaps the most common form in which the equivalence mentioned above is studied in the literature. In particular, the presence of chaos is proven by showing how a geometric and dynamic object similar in spirit to the horseshoe can be embedded into the flow of the period map. We have chosen to refrain from the standard horseshoe construction in this chapter, since it is thoroughly described in other sources. For example, Moser[2] contains a detailed account of the shift map and the existence of a horseshoe in the presence of transversal intersections of the stable and unstable manifolds. Similar discussions can be found in Guckenheimer & Holmes[3], and Wiggins[4].

The construction effected in the present chapter has been chosen for two reason. On the one hand, it offers an alternative to the standard horseshoe implementation and thus provides additional insight into the origin of chaos. On the other, the methods it employs will be important in later chapters. In particular, the emphasis on constructing shadowing orbits which jump between segments of the homoclinic orbit naturally leads to the idea of transition chains and Arnold diffusion discussed in Chapter 6. For a more extensive presentation of the ideas of exponential splittings, or dichotomies as they are often known in the literature, as well as proofs of many of the claims made in this chapter, we refer to Palmer[5].

Finally, for a discussion of Liapunov exponents and their relation to attractor dimensionality, albeit generalized to infinite dimensions, we suggest that the reader consult Temam[6].

1. S. Smale, "Diffeomorphisms with many periodic points," in *Differential and Combinatorial Topology*, ed. S. S. Cairns, pp. 63-80 (Princeton University Press, 1963).
2. J. Moser, *Stable and Random Motions in Dynamical Systems. With Special Emphasis on Celestial Mechanics* (Princeton University Press, 1973).
3. J. Guckenheimer and P. Holmes, *Nonlinear Oscillations, Dynamical Systems, and Bifurcations of Vector Fields* (Springer-Verlag, 1990).
4. S. Wiggins, *Global Bifurcations and Chaos: Analytical Methods* (Springer-Verlag, 1988).
5. K. J. Palmer, "Exponential Dichotomies and Transversal Homoclinic Points," *Journal of Differential Equations* **55** (1984), pp. 225-256.
6. R. Temam, *Infinite-Dimensional Dynamical Systems in Mechanics and Physics* (Springer-Verlag, 1988).

Chapter 4
THE PERTURBATION APPROACH - 1 $\frac{1}{2}$ DEGREES-OF-FREEDOM

As *real-politik* brings the meanderings of abstract political philosophy to the realm of everyday experiences, so applied mathematics attempts to accommodate the particularities of actual physical problems within a more abstract theory. The apparent success of this branch of mathematics, from the ancient geometers to today's inquiries into the interactions of exotic elementary particles, is undeniable. The underlying reasons for this amazing agreement between the processes of the physical world and the predictions of sometimes very unphysical theories continue to be a source of discourse and disagreement. A fascinating subject in itself, this text does not concern itself with these issues.

But a beautiful mathematical theory needs hooks to which actual applications can be attached. The applied mathematician desires devices that can detect the features responsible for the chaos and stochastic dynamics described in previous chapters. Given a dynamical model of a physical system, a most fruitful approach has been to seek another, simpler, dynamical system from which the first deviates only slightly. The analysis of the original system as a perturbation from the simpler one lies at the heart of **perturbation theory**.

Characteristic of the perturbation approach is

- that it relies on the simpler system being sufficiently understood, and
- that the conclusions about the perturbed system generally hold only for rather small perturbations; i.e., they are local in the space of dynamical systems.

In the context of Hamiltonian systems, the simpler system is generally integrable and thus completely understood. More generally, the simpler system possesses some symmetry, which is broken under the perturbation. Thus, in the Hamiltonian case, the perturbation method can at best give information on near-integrable systems such as those studied in connection with the KAM results in Chapter 2. Contrary to the KAM theorem, however, the discussion of the present chapter is on practical methods which, while less general than those of the KAM theory, allow simple implementation in particular applications.

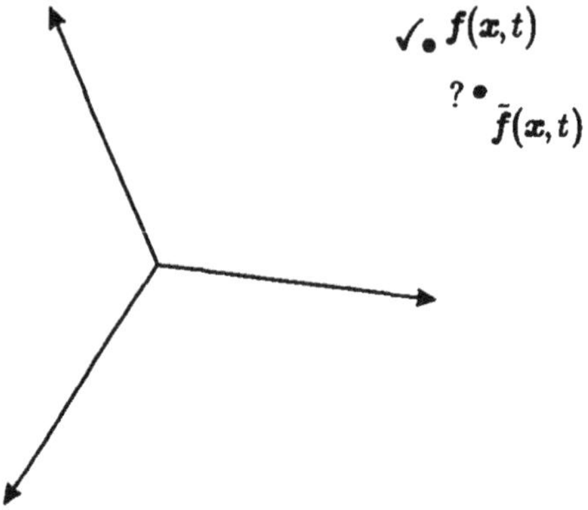

Perturbation theory bases its analysis of unknown systems on the known dynamics of nearby systems.

In this chapter we attempt to derive a perturbation method for the detection of transversal intersections of the stable and unstable manifolds of a periodic orbit. In Section 1 we achieve this by developing solutions on the perturbed manifolds as series expansions in the perturbation parameter with the unperturbed manifolds as zeroth-order terms. The complete solution of the series problem allows us to measure the separation of the manifolds and, consequently, determine whether an intersection has occurred and if it is transversal.

We apply the theoretical findings to a sample problem derived from the celestial mechanics application we shall consider in the next chapter. In particular, the problem is found to exhibit an additional level of degeneracy in that the perturbed manifolds separate only to second order in the perturbation.

4.1 Detecting transversal intersections

4.1.1 General properties

From previous chapters, we know that autonomous single-degree-of-freedom Hamiltonian systems are integrable and hence have very predictable dynamics. It was further noted that all hyperbolic equilibria have to be saddles with corresponding one-dimensional stable and unstable manifolds. An intersection of these manifolds would imply that the manifolds were in the same energy manifold. Since these are also one-dimensional, the manifolds would thus coincide to form a homo/heteroclinic orbit.

When two manifolds coincide in this way in a general dynamical system, the result is a highly structurally unstable situation. Most small perturbations of the dynamical system, while leaving the presence of the saddles and their stable and unstable manifolds unscathed, would lead to the separation of the manifolds. In the

Hamiltonian case, only if the perturbed system was effectively higher-dimensional could there be points of intersection left behind during separation. In the previous chapter we discussed the consequences of transversal intersections on the dynamics of the perturbed system. In this chapter we focus on developing a perturbation method for determining whether such intersections exist.

In particular, consider a periodically perturbed Hamiltonian system

$$\dot{q} = \frac{\partial H}{\partial p}(q,p,t;\varepsilon),\ \dot{p} = -\frac{\partial H}{\partial q}(q,p,t;\varepsilon), \quad q,p \in \mathcal{R}, \tag{4.1}$$

where the nonconservative Hamiltonian H is analytic in q, p, t and ε and

$$H(q,p,t;\epsilon) = H_0(q,p) + \varepsilon\tilde{H}(q,p,t;\varepsilon). \tag{4.2}$$

Here the perturbation $\tilde{H}$ is periodic with period T. The unperturbed system $\varepsilon = 0$ is further assumed to contain a homoclinic orbit, $\boldsymbol{x}_0(t)$, bi-asymptotic to the saddle point ζ_0.

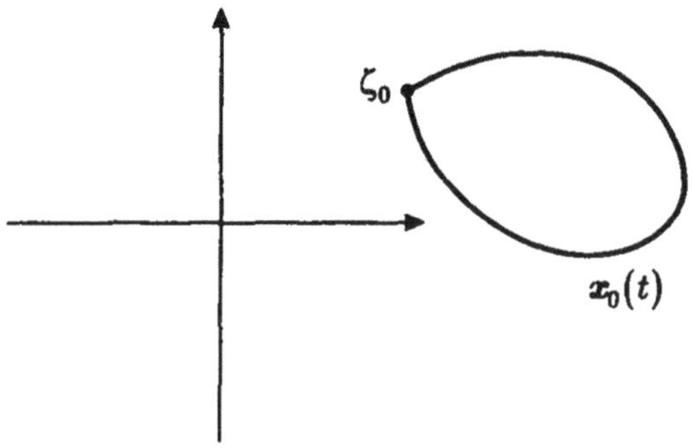

An orbit homoclinic to a saddle point.

Illustration

Our discussion of autonomous versus nonautonomous systems and the introduction of a suspended Hamiltonian in previous chapters suggest that the $1\frac{1}{2}$-degree-of-freedom system is equivalent to a certain 2-degree-of-freedom Hamiltonian. Conversely, a Hamiltonian dynamical system with Hamiltonian

$$H_\varepsilon(q,p,\varphi,I) = H_0(q,p,I) + \varepsilon H_1(q,p,\varphi,I) \tag{4.3}$$

such that

$$\omega(q,p,I) = \frac{\partial H_0}{\partial I}(q,p,I) > 0 \tag{4.4}$$

can be reduced to the form (4.1), at least for short times. In fact, since H_ε is conserved, we can write

$$H_0 + \varepsilon H_1 = h = \text{constant}. \tag{4.5}$$

Differentiating with respect to I at $\varepsilon = 0$, $I = I_0$ then yields $\omega(q, p, I_0)$. Since this is positive, it follows that, for I in an interval around I_0 and sufficiently small ε, Eq. (4.5) can be solved for I. For the same reason, the inversion is globally true in any compact subset of state space. If we write

$$I = -L(q, p, \varphi; \varepsilon, h) = -\sum_{k=0} \varepsilon^k L_k(q, p, \varphi; h) \tag{4.6}$$

and substitute this into Eq. (4.5), we obtain to zeroth order in ε

$$h = H_0(q, p, -L_0(q, p; h)), \tag{4.7}$$

which can be solved for L_0. Similarly, the ε and ε^2 terms yield

$$L_1(q, p, \varphi; h) = \frac{H_1(q, p, \varphi, -L_0(q, p; h)}{\omega(q, p, -L_0(q, p; h))} \tag{4.8}$$

and

$$L_2(q, p, \varphi; h) = -\frac{1}{2\omega(q, p, -L_0(q, p; h))} \frac{\partial}{\partial I} \left(\frac{(H_1(q, p, \varphi, I))^2}{\omega(q, p, I)} \right)_{I = -L_0(q, p; h)} \tag{4.9}$$

With the aim of applying the rescaling

$$\frac{dt}{ds} = \left(\frac{\partial H_\varepsilon}{\partial I}(q, p, \varphi, I) \right)^{-1}, \tag{4.10}$$

where the right-hand side is assumed to be strictly positive, we introduce the suspended Hamiltonian $H_\varepsilon^s = H_\varepsilon + p_0$, such that $H_\varepsilon^s \equiv 0$ on orbits. Using this, we see that

$$\varphi' = 1 \tag{4.11}$$

where $'$ denotes differentiation with respect to s. Thus, we can identify φ with s. Moreover,

$$q' = \frac{\partial H_\varepsilon}{\partial p} / \frac{\partial H_\varepsilon}{\partial I}, \text{ and } p' = -\frac{\partial H_\varepsilon}{\partial q} / \frac{\partial H_\varepsilon}{\partial I}. \tag{4.12}$$

But implicit differentiation of $H_\varepsilon(q, p, \varphi, -L(q, p, \varphi; \varepsilon, h) = h$ yields

$$\frac{\partial H_\varepsilon}{\partial (q, p)} - \frac{\partial H_\varepsilon}{\partial I} \frac{\partial L}{\partial (q, p)} = 0 \tag{4.13}$$

so that, finally, we see that q and p are canonically paired with respect to the independent variable φ through the periodic Hamiltonian $L(q, p, \varphi; \varepsilon, h)$. •

The period map $f_T(\cdot;0)$ of the unperturbed system clearly has a fixed point at ζ_0, such that the linearization $Df_T(\zeta_0;0)$ has no eigenvalues on the unit circle. It follows that $I - Df_T(\zeta_0;0)$ is invertible and, using the implicit function theorem, this implies the existence of an $\varepsilon_0 \geq 0$ such that the perturbed period map $f_T(\cdot\,;\varepsilon)$ has a unique, hyperbolic fixed point ζ_ε for $\varepsilon < \varepsilon_0$ in a neighborhood of ζ_0. Correspondingly, there is a hyperbolic periodic orbit, $\zeta_\varepsilon(t)$ of period T of the perturbed flow, with stable and unstable manifolds $\mathcal{W}^{s,u}(\zeta_\varepsilon(t))$.

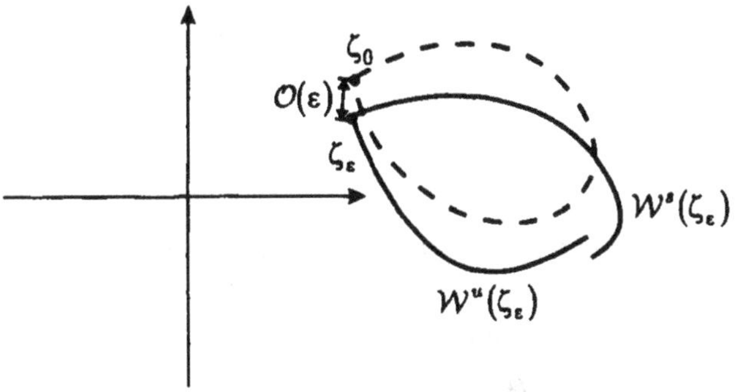

The stable and unstable manifolds survive the perturbation, but may no longer intersect.

It is actually possible to show that solutions on the perturbed manifolds can be expanded in power series in ε such that the zeroth-order term corresponds to motion on the unperturbed manifolds. In particular, consider a point ξ on the original homoclinic connection and denote by $x_0(t - t_0)$ the unperturbed solution for which $x_0(0) = \xi$. Then there exist unique solutions $x_\varepsilon^{s,u}(t,t_0)$ on the perturbed manifolds such that

$$x_\varepsilon^{s,u}(t,t_0) = x_0(t - t_0) + \sum_{n=1}^{\infty} \varepsilon^n x_n^{s,u}(t,t_0) \tag{4.14}$$

for $t \in [t_0,\infty)$ and $t \in (-\infty,t_0]$, respectively, for ε sufficiently small, and such that $(x_\varepsilon^{s,u}(t_0,t_0) - \xi)$ lies along the normal to the homoclinic manifold at ξ. While existence requires an involved and technical discussion, uniqueness is obtained by actual construction.

For convenience we introduce the normal

$$v = \left(\frac{\partial H^0}{\partial q}(\xi), \frac{\partial H^0}{\partial p}(\xi)\right)^T \tag{4.15}$$

along which we will eventually be interested in measuring the distance between $x_\varepsilon^s(t_0,t_0)$ and $x_\varepsilon^u(t_0,t_0)$. If we substitute the power series expansions into the equations of motion and expand in powers of ε we obtain the **variational equations**:

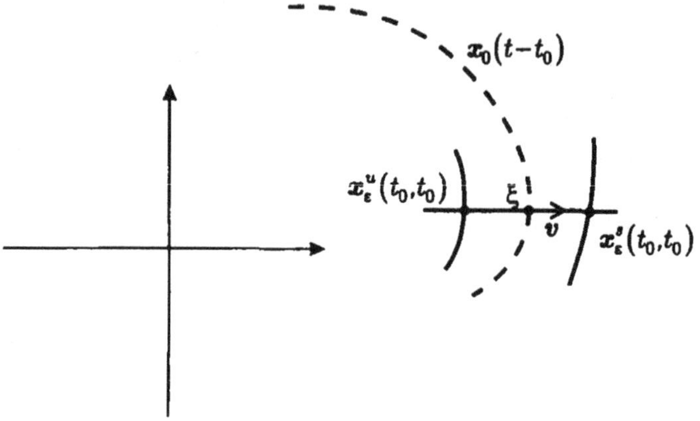

The perturbed manifolds intersect the normal in the points $x_\varepsilon^{s,u}(t_0, t_0)$.

$$\dot{x}_n^{s,u}(t,t_0) = JD^2H_0(x_0(t-t_0))x_n^{s,u}(t,t_0) + g_n(t, x_0(t-t_0), x_1^{s,u}(t,t_0), \ldots, x_{n-1}^{s,u}(t,t_0)) \tag{4.16}$$

to order n in ε. Here, the nonhomogeneity g_n is a known function of all lower-order terms and the resulting set of equations can then be solved iteratively.

4.1.2 Solving the variational equations

Since Eqs. (4.16) are linear, their solution is conveniently separated into finding a complete set of homogeneous solutions and a particular solution. It is straightforward to show that

$$(\phi_{11}(t-t_0), \phi_{12}(t-t_0))^T = \left(\frac{\partial H_0}{\partial p}(x_0(t-t_0)), -\frac{\partial H_0}{\partial q}(x_0(t-t_0))\right)^T \tag{4.17}$$

solves the homogeneous part of Eq. (4.16). Over intervals $\mathcal{I}$ such that $\phi_{11}(t-t_0) \neq 0$ for $t \in \mathcal{I}$, a second homogeneous solution is given by $(\phi_{21}(t-t_0), \phi_{22}(t-t_0))^T$, where

$$\phi_{21}(t-t_0) = \phi_{11}(t-t_0)\int_{t_0}^{t} \frac{\partial^2 H_0}{\partial p^2}(x_0(s-t_0))\,[\phi_{11}(s-t_0)]^{-2}\,ds \tag{4.18}$$

and

$$\phi_{22}(t-t_0) = \phi_{12}(t-t_0)\int_{t_0}^{t} \frac{\partial^2 H_0}{\partial p^2}(x_0(s-t_0))\,[\phi_{11}(s-t_0)]^{-2}\,ds + [\phi_{11}(t-t_0)]^{-1}. \tag{4.19}$$

Similarly, over intervals such that $\phi_{12} \neq 0$, an alternative expression is given by

$$\phi_{21}(t-t_0) = \phi_{11}(t-t_0)\int_{t_0}^{t} \frac{\partial^2 H_0}{\partial q^2}(x_0(s-t_0))\,[\phi_{12}(s-t_0)]^{-2}\,ds - [\phi_{12}(t-t_0)]^{-1} \tag{4.20}$$

and

$$\phi_{22}(t-t_0) = \phi_{12}(t-t_0)\int_{t_0}^{t} \frac{\partial^2 H_0}{\partial q^2}(\boldsymbol{x}_0(s-t_0))\,[\phi_{12}(s-t_0)]^{-2}\,ds. \tag{4.21}$$

In particular, the Wronskian

$$\phi_{11}\phi_{22} - \phi_{21}\phi_{12} \equiv 1 \tag{4.22}$$

for all times. If necessary, the second homogeneous solution could be obtained for $t \in (-\infty, \infty)$ by patching together solutions of the two types above. Clearly, ϕ_{11} and ϕ_{12} do not vanish simultaneously other than at an equilibrium, and thus not on the homoclinic manifold.

Illustration

We introduce the fundamental matrix

$$\boldsymbol{X}(t,t_0) = \begin{pmatrix} \phi_{11}(t-t_0) & \phi_{21}(t-t_0) \\ \phi_{12}(t-t_0) & \phi_{22}(t-t_0) \end{pmatrix} \tag{4.23}$$

for the homogeneous part of Eq. (4.16). Furthermore, the projection matrix

$$\boldsymbol{P}(t,t_0) = \begin{pmatrix} \phi_{11}(t-t_0)\phi_{22}(t-t_0) & -\phi_{11}(t-t_0)\phi_{21}(t-t_0) \\ \phi_{12}(t-t_0)\phi_{22}(t-t_0) & -\phi_{12}(t-t_0)\phi_{21}(t-t_0) \end{pmatrix} \tag{4.24}$$

maps $(\phi_{11}(t-t_0), \phi_{12}(t-t_0))^T$ *to itself and* $(\phi_{21}(t-t_0), \phi_{22}(t-t_0))^T$ *to* $\mathbf{0}$.

Since the former vector spans the tangent space to the unperturbed homoclinic manifold, the results of previous chapters imply

$$|\boldsymbol{X}(t,t_0)\boldsymbol{X}^{-1}(s,t_0)\boldsymbol{P}(s,t_0)| \le Ke^{-\lambda(t-s)},\ t \ge s \ge t_0 \tag{4.25}$$

and

$$|\boldsymbol{X}(t,t_0)\boldsymbol{X}^{-1}(s,t_0)(\boldsymbol{I} - \boldsymbol{P}(s,t_0))| \le Ke^{-\lambda(s-t)},\ s \ge t \ge t_0. \tag{4.26}$$

Writing out the operators in matrix form yields

$$\begin{pmatrix} \phi_{11}(t-t_0)\phi_{22}(s-t_0) & -\phi_{11}(t-t_0)\phi_{21}(s-t_0) \\ \phi_{12}(t-t_0)\phi_{22}(s-t_0) & -\phi_{12}(t-t_0)\phi_{21}(s-t_0) \end{pmatrix} \tag{4.27}$$

and

$$\begin{pmatrix} -\phi_{21}(t-t_0)\phi_{12}(s-t_0) & \phi_{21}(t-t_0)\phi_{11}(s-t_0) \\ -\phi_{22}(t-t_0)\phi_{12}(s-t_0) & \phi_{22}(t-t_0)\phi_{11}(s-t_0) \end{pmatrix}, \tag{4.28}$$

and bounds like those above (but with different K*) are valid for each component by equivalence of norms.*

Similar bounds can also be found for $s, t \le t_0$ *by following the homoclinic manifold backwards in time toward* ζ. •

A particular solution is then simply obtained through the variation of parameters method

$$\begin{pmatrix} \phi^p_{n1}(t,t_0) \\ \phi^p_{n2}(t,t_0) \end{pmatrix} = \boldsymbol{X}(t,t_0)\int_{t_0}^{t} \boldsymbol{X}^{-1}(s,t_0)\boldsymbol{g}_n(s,t_0)ds \tag{4.29}$$

where we have used the simplified notation

$$\boldsymbol{g}_n(s,t_0) = \boldsymbol{g}_n\left(s, \boldsymbol{x}_0(s-t_0), \boldsymbol{x}_1(s,t_0), \ldots, \boldsymbol{x}_{n-1}(s,t_0)\right).$$

Putting all this together yields the general solution to the n-th variational equation

$$\boldsymbol{x}_n(t,t_0) = \boldsymbol{X}(t,t_0)\left[\begin{pmatrix} \alpha_n(t_0) \\ \beta_n(t_0) \end{pmatrix} + \int_{t_0}^{t} \boldsymbol{X}^{-1}(s,t_0)\boldsymbol{g}_n(s,t_0)ds\right] \tag{4.30}$$

where $\alpha_n(t_0)$, $\beta_n(t_0)$ are arbitrary constants parametrized by t_0. The condition that $(\boldsymbol{x}^{s,u}_\varepsilon(t_0,t_0) - \boldsymbol{\xi})$ be perpendicular to the unperturbed homoclinic manifold translates into

$$\left(\phi_{11}^2(0) + \phi_{12}^2(0), \phi_{11}(0)\phi_{21}(0) + \phi_{12}(0)\phi_{22}(0)\right)\begin{pmatrix} \alpha_n^{s,u}(t_0) \\ \beta_n^{s,u}(t_0) \end{pmatrix} = 0 \tag{4.31}$$

to ε^n. Since $\phi_{11}^2(0) + \phi_{12}^2(0) \neq 0$, it follows that $\alpha_n^{s,u}(t_0)$ is uniquely determined once $\beta_n^{s,u}(t_0)$ is fixed.

From the bounds found in the illustration, it follows that the terms containing $\phi_{11}(t-t_0)$ and $\phi_{12}(t-t_0)$ in (4.30) are bounded as $|t| \to \infty$. This is not so, however, for those containing $\phi_{21}(t-t_0)$ and $\phi_{22}(t-t_0)$ for general $\beta_n(t_0)$. If we write the coefficient of $\phi_{21}(t-t_0)$ (and $\phi_{22}(t-t_0)$) as $\beta_n(t_0) - \int_{t_0}^{t} c_n(s,t_0)ds$, then only if

$$\beta_n^{s,u}(t_0) = \int_{t_0}^{\infty,-\infty} c_n(s,t_0)ds \tag{4.32}$$

so that the remaining integral is

$$\int_{t}^{\infty,-\infty} c_n(s,t_0)ds \tag{4.33}$$

do they remain bounded. This choice then uniquely fixes $\beta_n(t_0)$ and, consequently, $\alpha_n(t_0)$, which was what we set out to prove.

4.1.3 Separation of the manifolds

It remains to determine whether the perturbed manifolds intersect and, if so, if the intersection is transversal. An obvious candidate measure for the distance between the manifolds is given by

$$d(t_0;\varepsilon) = \boldsymbol{v}\cdot(\boldsymbol{x}_\varepsilon^u(t_0,t_0) - \boldsymbol{x}_\varepsilon^s(t_0,t_0))$$

$$= \boldsymbol{v}\cdot\sum_{n=1}^{\infty}\varepsilon^n \boldsymbol{X}(t_0,t_0)\left[\begin{pmatrix}\alpha_n^u(t_0)\\ \beta_n^u(t_0)\end{pmatrix} - \begin{pmatrix}\alpha_n^s(t_0)\\ \beta_n^s(t_0)\end{pmatrix}\right] = \sum_{n=1}^{\infty}\varepsilon^n d_n(t_0) \tag{4.34}$$

where we have expressed $\alpha_n(t_0)$ in terms of $\beta_n(t_0)$ using Eq. (4.31) and $d_n(t_0) = \beta_n^u(t_0) - \beta_n^s(t_0)$. Clearly, d_n is simply evaluated by solving the variational equations for $k \leq n-1$. A zero of $d(t_0;\varepsilon)$ corresponds to an intersection of the perturbed manifolds. In fact, if $d_1(\tau) = 0$ and $d_1'(\tau) \neq 0$, then the implicit function theorem implies that there is a unique continuous function $\tau^*(\varepsilon)$ such that $d(\tau^*(\varepsilon);\varepsilon) = 0$ for sufficiently small ε, i.e., an intersection can be detected by simply examining d_1. It also follows that $d'(\tau^*(\varepsilon);\varepsilon) \neq 0$, i.e., the intersection is transversal. A similar conclusion follows by studying d_l in the case that $d_k \equiv 0$ for $k < l$.

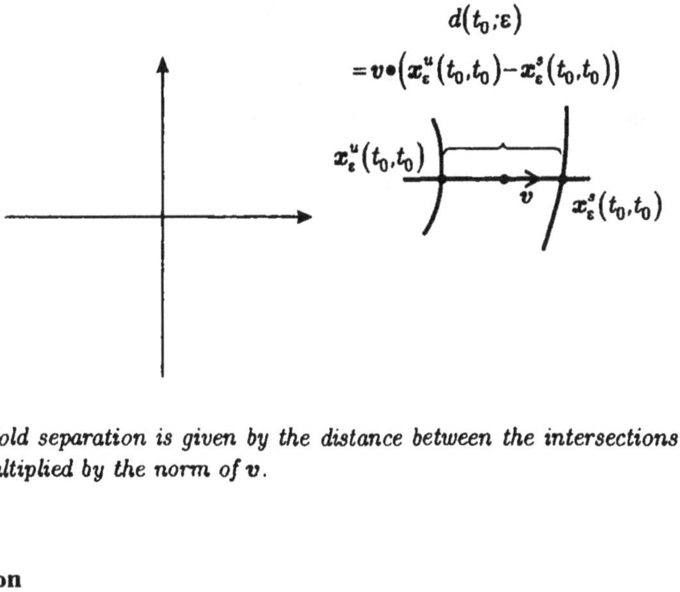

The manifold separation is given by the distance between the intersections with the normal multiplied by the norm of $\boldsymbol{v}$.

Illustration

From Eq. (4.32) we find

$$d_n(t_0) = \int_{t_0}^{\infty} c_n^s(s,t_0)ds + \int_{-\infty}^{t_0} c_n^u(s,t_0)ds. \tag{4.35}$$

Clearly, the integrands are identical if $\boldsymbol{x}_k^s \equiv \boldsymbol{x}_k^u$ for $k \leq n-1$, in which case we can write

$$d_n(t_0) = \int_{-\infty}^{\infty} c_n(s,t_0)ds. \tag{4.36}$$

Expanding $\tilde{H}$ in Eq. (4.2) in a power series in ε, the integrands for $n = 1, 2$ are easily found to be the Poisson brackets $\{H_0, H_1\}$, and

$$\left\{H_0, \left[H_2 + \boldsymbol{x}_1(t, t_0) \cdot \boldsymbol{D}H_1 + \frac{1}{2}\boldsymbol{x}_1(t, t_0) \cdot \boldsymbol{D}^2 H_0 \boldsymbol{x}_1(t, t_0)\right]\right\} \tag{4.37}$$

respectively, where the Poisson brackets are evaluated along the unperturbed homoclinic solution $\boldsymbol{x}_0(t - t_0)$. It is hopefully clear how higher-order integrands could be generated. •

An alternative measure of the distance between the manifolds is given by

$$\tilde{d}(t_0; \varepsilon) = H(\boldsymbol{x}_\varepsilon^u(t_0, t_0), t_0; \varepsilon) - H(\boldsymbol{x}_\varepsilon^s(t_0, t_0), t_0; \varepsilon) \tag{4.38}$$

i.e., the difference in energy between the two points on the perturbed manifolds. In fact,

$$\tilde{d}(t_0; \varepsilon) = \boldsymbol{D}H(\boldsymbol{\xi}', t_0; \varepsilon) \cdot (\boldsymbol{x}_\varepsilon^u(t_0, t_0) - \boldsymbol{x}_\varepsilon^s(t_0, t_0)), \tag{4.39}$$

where $\boldsymbol{\xi}'$ is some point on the line segment between $\boldsymbol{x}_\varepsilon^s(t_0, t_0))$ and $\boldsymbol{x}_\varepsilon^u(t_0, t_0))$. However, the gradient $\boldsymbol{D}H$ is a small perturbation of the gradient of H_0 which is nonzero in a neighborhood of ξ and directed along the normal $\boldsymbol{v}$. Thus, for sufficiently small ϵ, $\boldsymbol{D}H(\boldsymbol{\xi}', t_0; \varepsilon)$ has a nonzero component in the direction of $\boldsymbol{v}$. Consequently, simple zeroes of $\tilde{d}$ correspond in a one-to-one fashion to transversal intersections of the perturbed manifolds.

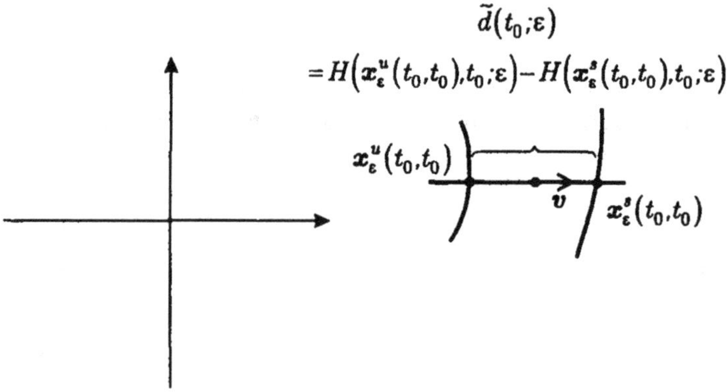

The manifold separation is given by the difference in energy of the intersections with the normal.

To evaluate $\tilde{d}$ to, say, ε^n simply from its definition requires solving the variational equations for $k \leq n$, i.e. one more than in the case of d. In fact, an improvement is possible from the following observation. From previous chapters we have

$$\frac{dH}{dt} = \varepsilon \frac{\partial \tilde{H}}{\partial t}, \tag{4.40}$$

which implies that

$$H(\boldsymbol{x}_\varepsilon(t,t_0),t;\varepsilon) - H(\boldsymbol{x}_\varepsilon(t_0,t_0),t_0;\varepsilon) = \varepsilon \int_{t_0}^{t} \frac{\partial \tilde{H}}{\partial t}(\boldsymbol{x}_\varepsilon(s,t_0),s;\varepsilon)\mathrm{d}s. \tag{4.41}$$

Both sides of this equation approach periodic functions as $t \to \pm\infty$, on the stable and unstable manifolds, respectively. Hence, the zeroth Fourier components of these functions must agree. In the homoclinic case, the left-hand side approaches the same periodic function as $|t| \to \infty$, and we thus find

$$\tilde{d}(t_0;\epsilon) = \varepsilon \overline{\int_{t_0}^{t\to\infty} \frac{\partial \tilde{H}}{\partial t}(\boldsymbol{x}_\varepsilon(s,t_0),s;\varepsilon)\mathrm{d}s} + \varepsilon \overline{\int_{t\to-\infty}^{t_0} \frac{\partial \tilde{H}}{\partial t}(\boldsymbol{x}_\varepsilon(s,t_0),s;\varepsilon)\mathrm{d}s}, \tag{4.42}$$

where the bar notation refers to the mean value of the integral as $t \to \pm\infty$, respectively. It is easy to see that evaluation of $\tilde{d}$ to order n now requires solving only the variational equations for $k \le n-1$.

Illustration

To first order in ε, the relevant integrand is

$$\frac{\partial H_1}{\partial t}(\boldsymbol{x}_0(s-t_0),s) = \frac{dH_1}{dt}(\boldsymbol{x}_0(s-t_0),s) + \{H_0,H_1\}(\boldsymbol{x}_0(s-t_0),s). \tag{4.43}$$

Integrating over $[t_0,t]$ yields

$$\int_{t_0}^{t} \frac{\partial H_1}{\partial t}(\boldsymbol{x}_0(s-t_0),s)\mathrm{d}s$$

$$= H_1(\boldsymbol{x}_0(t-t_0),t) - H_1(\boldsymbol{\xi},t_0) + \int_{t_0}^{t} \{H_0,H_1\}(\boldsymbol{x}_0(s-t_0),s)\mathrm{d}s. \tag{4.44}$$

As $t \to \pm\infty$, the first term on the right-hand side approaches the same periodic function. Similarly, the second term is independent of whether we are on the stable or unstable manifold. Finally, the integral is absolutely convergent. Hence

$$\tilde{d}(t_0;\varepsilon) = \varepsilon \int_{-\infty}^{\infty} \{H_0,H_1\}(\boldsymbol{x}_0(s-t_0),s)\mathrm{d}s + \mathcal{O}(\varepsilon^2) \tag{4.45}$$

i.e., identical to $d(t_0;\varepsilon)$ to $\mathcal{O}(\varepsilon^2)$. To ε^2 the integrand is

$$\boldsymbol{x}_1(s,t_0)\cdot\frac{\partial^2 H_1}{\partial t\partial \boldsymbol{x}}(\boldsymbol{x}_0(s-t_0),s) + \frac{\partial H_2}{\partial t}(\boldsymbol{x}_0(s-t_0),s), \tag{4.46}$$

where the last term results in the absolutely convergent integral

$$\int_{-\infty}^{\infty}\{H_0, H_2\}(x_0(s-t_0), s)ds. \tag{4.47}$$

•

From the illustrations it is clear that the integrands in the evaluation of $\tilde{d}$ are algebraically simpler than those for d. On the other hand, the former requires the determination of the asymptotic mean, whereas the latter involves only absolutely convergent integrals. It is easy to show that the two yield identical numerical answers in an order n calculation, provided that the manifolds coincide to order $n-1$. Since higher-order calculations usually require numerical evaluation, this provides a convenient means for checking one's results.

In the heteroclinic case, Eq. (4.42) does not suffice. It becomes necessary to include the first term on the left-hand side of Eq. (4.41). When, as often is the case, the heteroclinic connection is the result of a particular symmetry, which in turn is respected by the perturbation, this term is again unimportant.

4.2 A relevant sample problem

We consider the Hamiltonian system generated by the Hamiltonian

$$H = H_0(q,p) + \varepsilon H_1(q,p,t) = \frac{1}{8}p^2 + 2q^2 - q^4$$

$$+\varepsilon 8\sqrt{2}k\sqrt{h}\,\mathrm{sn}\left[h^{\frac{1}{4}}t,k\right]\mathrm{dn}\left[h^{\frac{1}{4}}t,k\right] q\left(\frac{1}{3}q^2 + 2k^2\sqrt{h}\,\mathrm{cn}^2\left[h^{\frac{1}{4}}t,k\right]\right), \tag{4.48}$$

where sn, cn, and dn are Jacobian elliptic functions; k is the modulus; and we assume that

$$k^2 = \frac{\sqrt{h}-1}{2\sqrt{h}}, \; h > 0. \tag{4.49}$$

The elliptic functions are periodic with period $4K(k)$, where K is a complete elliptic integral. The rather complicated perturbation has been chosen to produce similar calculations to those we will encounter in the next chapter when considering a specific celestial mechanics problem. In particular, d_1 is identical in the two cases. The relation (4.49) between k and h is another reflection of the physical problem where the two quantities have particular physical meaning.

The phase portrait of the unperturbed system displays the existence of two heteroclinic connections between $(\pm 1, 0)$ to form a heteroclinic cycle. We expect the perturbation to break the connections and leave behind points of transversal intersections between the perturbed manifolds. The results of the previous chapter then imply the existence of solutions that jump in a seemingly random fashion between the two hyperbolic periodic orbits. We shall apply the theory of the preceding section to actually prove the existence of transversal homoclinic points.

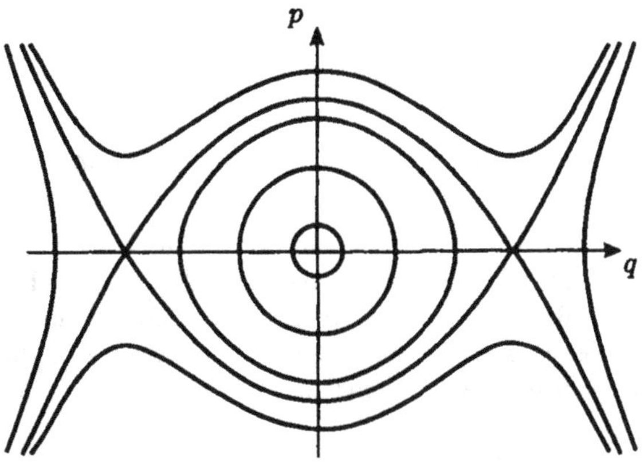

The heteroclinic cycle formed by the two heteroclinic orbits.

We note that H is unchanged under the reflection $(q, t) \to (-q, -t)$. It thus follows that if $(q_+(t), p_+(t))$ is a solution to the perturbed system, then indeed so is $(q_-(t), p_-(t)) = (-q_+(-t), p_+(-t))$. Consequently, the two hyperbolic periodic orbits near the original saddle points are simply each other's mirror images traversed in the same direction. It further follows that H evaluated on these periodic orbits yields identical periodic functions and, consequently, the homoclinic theory applies (cf. Eq. (4.42)).

Let $\boldsymbol{\xi}_\pm = (0, \pm 2\sqrt{2})$ be the highest and lowest point on the upper (+) and lower (-) homoclinic connection, respectively. Then

$$\boldsymbol{x}^T_{0,+}(t - t_0) = \left(\pm\tanh\left(\frac{t-t_0}{\sqrt{2}}\right), \pm 2\sqrt{2}\mathrm{sech}^2\left(\frac{t-t_0}{\sqrt{2}}\right)\right). \tag{4.50}$$

We are now ready to evaluate the manifold separation to ε:

$$d_{1,\pm}(t_0) = \int_{-\infty}^{\infty}\{H_0, H_1\}(\boldsymbol{x}_{0,\pm}(t-t_0), t)dt = \mp 8k\sqrt{h}\int_{-\infty}^{\infty} \mathrm{sn}\left[h^{\frac{1}{4}}t, k\right]\, \mathrm{dn}\left[h^{\frac{1}{4}}t, k\right]$$

$$\cdot\,\mathrm{sech}^2\left(\frac{t-t_0}{\sqrt{2}}\right)\left(\tanh^2\left(\frac{t-t_0}{\sqrt{2}}\right) + 2k^2\sqrt{h}\,\mathrm{cn}^2\left[h^{\frac{1}{4}}t, k\right]\right) dt. \tag{4.51}$$

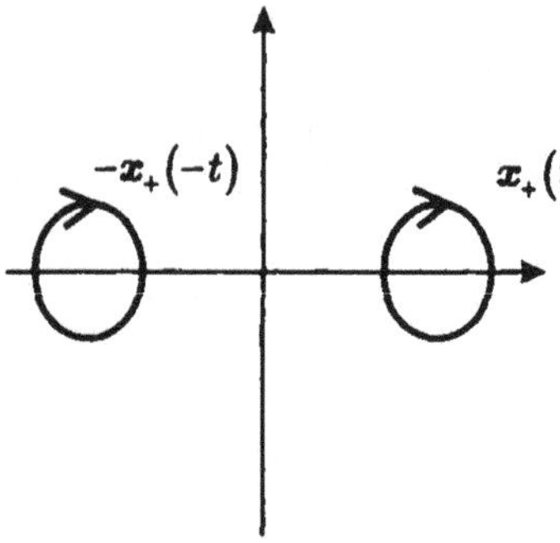

The perturbed periodic orbits are mirror images traversed in the same direction.

In a later section we evaluate this integral using residue calculations. Unexpectedly, for all h, k satisfying Eq. (4.49), one finds $d_{1,\pm}(t_0) \equiv 0$! In other words, the perturbed manifolds do not simply intersect along a homoclinic orbit, but in fact, coincide to first order in ε. In order to discern a separation of the manifolds, it is consequently necessary to proceed to higher order.

4.2.1 The first variational equation

A second-order calculation of $\tilde{d}$ requires the explicit knowledge of $\boldsymbol{x}_1$; i.e., we need to solve the first variational equation. The general solution is already given in Eq. (4.30), where

$$\boldsymbol{g}_1(s,t_0) = \begin{pmatrix} 0 \\ \gamma(s,t_0) \end{pmatrix}$$

$$= \begin{pmatrix} 0 \\ -8\sqrt{2}k\sqrt{h}\,\mathrm{sn}\left[h^{\frac{1}{4}}s,k\right]\mathrm{dn}\left[h^{\frac{1}{4}}s,k\right]\left(\tanh^2\left(\frac{s-t_0}{\sqrt{2}}\right)+2k^2\sqrt{h}\,\mathrm{cn}^2\left[h^{\frac{1}{4}}s,k\right]\right) \end{pmatrix} \tag{4.52}$$

and it remains to find $\boldsymbol{X}(t,t_0)$. Following previous sections we find

$$(\phi_{11,\pm}(t-t_0), \phi_{12,\pm}(t-t_0))$$

$$= \pm\left(\frac{1}{\sqrt{2}}\mathrm{sech}^2\left(\frac{t-t_0}{\sqrt{2}}\right), -4\tanh\left(\frac{t-t_0}{\sqrt{2}}\right)\mathrm{sech}^2\left(\frac{t-t_0}{\sqrt{2}}\right)\right) \tag{4.53}$$

and since $\phi_{11,\pm}(0) \neq 0$, we can apply Eqs. (4.18-4.19) to yield

$$\phi_{21,+}(t-t_0) = \frac{1}{16}\sinh\left(\sqrt{2}(t-t_0)\right)$$

$$+\frac{3}{16}\tanh\left(\frac{t-t_0}{\sqrt{2}}\right) + \frac{3\sqrt{2}}{32}(t-t_0)\,\mathrm{sech}^2\left(\frac{t-t_0}{\sqrt{2}}\right) \tag{4.54}$$

and

$$\phi_{22,+}(t-t_0) = \frac{1}{2\sqrt{2}}\cosh\left(\sqrt{2}(t-t_0)\right)$$

$$+\frac{3}{2\sqrt{2}}\operatorname{sech}^2\left(\frac{t-t_0}{\sqrt{2}}\right) - \frac{3}{4}(t-t_0)\tanh\left(\frac{t-t_0}{\sqrt{2}}\right)\operatorname{sech}^2\left(\frac{t-t_0}{\sqrt{2}}\right) \tag{4.55}$$

such that $\phi_{21,-} \equiv -\phi_{21,+}$ and $\phi_{22,-} \equiv -\phi_{22,+}$.

Since $\phi_{12,\pm}(0) = \phi_{21,\pm}(0) = 0$, it follows from Eq. (4.31) that $\alpha^{s,u}_{1,\pm}(t_0) \equiv 0$. Moreover, since $x^s_{1,\pm}(t,t_0) \equiv x^u_{1,\pm}(t,t_0)$, the theory in the previous sections implies that

$$\beta^{s,u}_{1,\pm}(t_0) = -\int_{t_0}^{\infty} \phi_{11,\pm}(s-t_0)\gamma(s,t_0)ds. \tag{4.56}$$

We finally obtain

$$x^{s,u}_{1,+}(t,t_0) = X_+(t,t_0)\begin{pmatrix} -\int_{t_0}^{t}\phi_{21,+}(s-t_0)\gamma(s,t_0)ds \\ -\int_{t}^{\infty}\phi_{11,+}(s-t_0)\gamma(s,t_0)ds \end{pmatrix} \tag{4.57}$$

and $x_{1,-}(t,t_0) \equiv -x_{1,+}(t,t_0)$. We remark that

$$q_{1,\pm}(t,0) = -q_{1,\pm}(-t,0) \text{ and } p_{1,\pm}(t,0) = p_{1,\pm}(-t,0). \tag{4.58}$$

4.2.2 Second-order calculations

We proceed to evaluate $\tilde{d}_{2,\pm}(t_0)$. From Eq. (4.46) we see that this involves finding the asymptotic mean of the function

$$\int_{t_0}^{t} q_{1,\pm}(s,t_0)\frac{\partial^2 H_1}{\partial t\partial q}(x_{0,\pm}(s-t_0),s)ds \tag{4.59}$$

as $t \to \pm\infty$, respectively. Here

$$\frac{\partial^2 H_1}{\partial t\partial q}(x_{0,\pm}(s-t_0),s) = 8\sqrt{2}kh^{\frac{3}{4}}\mathrm{cn}\left[(\mathrm{dn}^2 - k^2\mathrm{sn}^2)\right.$$

$$\left.\cdot\left(\tanh^2\left(\frac{s-t_0}{\sqrt{2}}\right) + 2k^2\sqrt{h}\mathrm{cn}^2\right) - 4k^2\sqrt{h}\mathrm{sn}^2\mathrm{dn}^2\right], \tag{4.60}$$

where we have omitted the $[h^{\frac{1}{4}}s,k]$ argument of the elliptic functions. Clearly, the result for the upper and lower homoclinic connections differ only in sign and, hence, we perform the calculations only for the upper connection, omitting the subscript.

For $t_0 = 0$, the integrand in Eq. (4.59) is odd and thus, the integral is even in t. It follows that the asymptotic means as $t \to \infty$ and $-\infty$ have to be equal, and consequently, $\tilde{d}_2(0) = 0$. To show that this zero is simple, we differentiate the integral (4.59) with respect to t_0 and subsequently set $t_0 = 0$. It is easy to show that in this case, the integrand is even and it thus suffices to calculate the asymptotic mean as $t \to \infty$. Doubling the result thus yields the rate of change of the separation between the manifolds. A numerical evaluation of the integrand for $h = 4$ shows that the integral is absolutely convergent and the asymptotic mean simply equals the value of the integral as $t \to \infty$. This is easily evaluated numerically, and we find $\tilde{d}_2'(0) \approx -1.3 \neq 0$. Thus, for $h = 4$ and, by analyticity, for all but a discrete number of points, the manifolds intersect transversely. The discussion in the previous chapter now applies to imply the conjugacy of the Bernoulli shift to an iterate of the period map. Moreover, we have shown the existence of shadowing orbits that move along segments of the original heteroclinic manifolds in a random way.

4.3 Evaluating the lowest-order separation

We will now show how to evaluate the integral in Equation (4.51)

$$d_{1,\pm}(t_0) = \mp 8k\sqrt{h} \int_{-\infty}^{\infty} \frac{\text{sn dn}}{c^2} \left[\frac{s^2}{c^2} + 2k^2\sqrt{h}\,\text{cn}^2\right] \text{d}t, \tag{4.61}$$

where we have used the abbreviations $\text{sn} = \text{sn}[h^{\frac{1}{4}}(t+t_0), k]$, $\text{dn} = \text{dn}[h^{\frac{1}{4}}(t+t_0), k]$, $\text{cn} = \text{cn}[h^{\frac{1}{4}}(t+t_0), k]$, $s = \sinh\frac{t}{\sqrt{2}}$ and $c = \cosh\frac{t}{\sqrt{2}}$. Note that we have performed a variable substitution $t \to t + t_0$ in the integral. Integrating by parts yields

$$d_1(t_0) = \pm 8k\sqrt{h} \left[\frac{\sqrt{2}}{h^{\frac{1}{4}}} \int_{-\infty}^{\infty} \frac{s^3 - s}{c^5}\,\text{cn}\,dt + \frac{2\sqrt{2}}{3} k^2 h^{\frac{1}{4}} \int_{-\infty}^{\infty} \frac{s}{c^3}\,\text{cn}^3 \text{d}t\right]. \tag{4.62}$$

By expanding the elliptic functions appearing in these integrals in Fourier series, we can rewrite the bracketed expression as

$$\left(\frac{\sqrt{2}}{h^{\frac{1}{4}}} \sum_{n=-\infty}^{\infty} a_n I_1(n) + \frac{2\sqrt{2}}{3} k^2 h^{\frac{1}{4}} \sum_{n=-\infty}^{\infty} b_n I_2(n)\right) \exp\left[\frac{in\pi h^{\frac{1}{4}} t_0}{2K(k)}\right], \tag{4.63}$$

where a_n and b_n are the Fourier coefficients of cn and cn^3, respectively. Here,

$$I_1(n) = \int_{-\infty}^{\infty} \frac{s^3 - s}{c^5} \exp\left[\frac{in\pi h^{\frac{1}{4}} t}{2K(k)}\right] dt \tag{4.64}$$

and

$$I_2(n) = \int_{-\infty}^{\infty} \frac{s}{c^3} \exp\left[\frac{in\pi h^{\frac{1}{4}} t}{2K(k)}\right] dt, \tag{4.65}$$

where $K(k)$ is the complete elliptic integral of the first kind. (Recall that the period of the Jacobian elliptic functions is $4K(k)$.) We evaluate these two integrals by residue integration over the closed contour shown below. The results of this procedure are

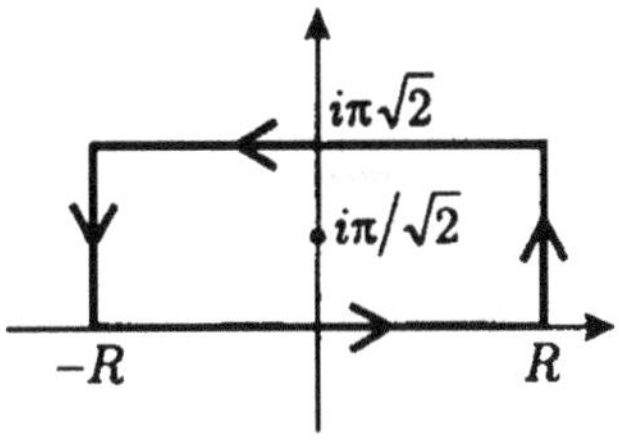

Contour integration as $R \to \infty$.

$$I_2(n) = \frac{2\sqrt{2}\pi i\Lambda}{1-\Lambda^2}\frac{n^2\pi^2\sqrt{h}}{4K^2(k)}, \text{ where } \Lambda = \exp\left[-\frac{n\pi^2 h^{\frac{1}{4}}}{2\sqrt{2}K(k)}\right] \tag{4.66}$$

and

$$I_1(n) = \frac{I_2(n)}{3}\left(1 - \frac{n^2\pi^2\sqrt{h}}{4K^2(k)}\right), \tag{4.67}$$

for $n \neq 0$ and $I_1 = I_2 = 0$ for $n = 0$.

We now need the Fourier coefficients of cn and cn^3. We again use contour integration. For example, a_n is given by

$$4K(k)a_n = \int_{K(k)}^{5K(k)} \mathrm{cn}[u,k]e^{-\frac{in\pi u}{2K(k)}}du, \tag{4.68}$$

evaluated by integrating over the closed contour shown below. Taylor expanding cn near $z = iK'(k)$ we find

$$\mathrm{cn}[z,k] = -\frac{i}{k}\left[\frac{1}{z - iK'(k)} + \frac{1-2k^2}{6}(z - iK'(k) + \cdots\right], \tag{4.69}$$

where $K'(k)$ denotes the complementary complete elliptic integral of the first kind. Furthermore, $\mathrm{cn}(z + 2K(k)) = -\mathrm{cn}(z)$ so that

$$\mathrm{cn}[z,k] = \frac{i}{k}\left[\frac{1}{z - 2K - iK'(k)} + \frac{1-2k^2}{6}(z - 2K - iK'(k) + \cdots\right]. \tag{4.70}$$

In other words, cn has simple poles at $iK'(k) + 4mK(k)$ and $\pm 2K(k) + 4mK(k) + iK'(k)$ with residues $-\frac{i}{k}$ and $\frac{i}{k}$, respectively, for every m. Using the residue theorem one then finds

$$a_n = \begin{cases} 0 & , \quad n = even \\ \frac{\pi}{kK(k)}\frac{q^{m+\frac{1}{2}}}{1+q^{2m+1}} & , \quad n = odd = 2m+1 \end{cases}, \tag{4.71}$$

where $q = exp(-\frac{\pi K'(k)}{K(k)})$ is the elliptic nome.

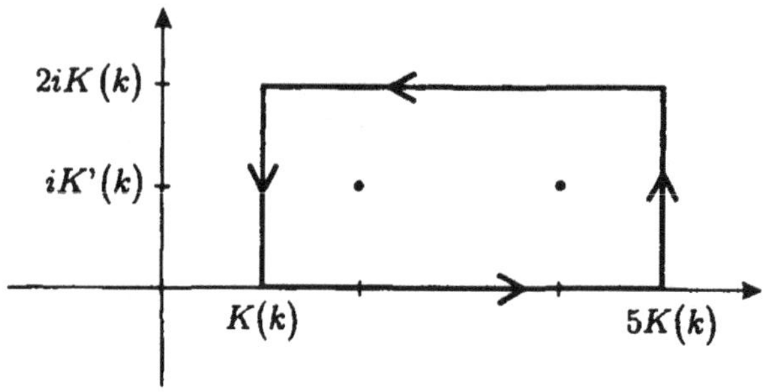

Contour integration for the Fourier coefficients of the doubly periodic elliptic functions.

From the Taylor expansion we obtain the residues of $\mathrm{cn}^3(z)\exp\left[-\frac{in\pi z}{2K(k)}\right]$ at the poles $iK'(k)$ and $\pm 2K(k)+iK'(k)$. Again, using the same contour and the residue theorem, we obtain

$$b_n = \begin{cases} 0 & , \quad n = even \\ \frac{\pi}{8K^3(k)k^3}[\pi^2 n^2 + 8K^2(k)k^2 - 4K^2(k)]\frac{q^{m+\frac{1}{2}}}{1+q^{2m+1}} & , \quad n = 2m+1 \end{cases} . \tag{4.72}$$

Substituting all these results into our expression for $d_{1,\pm}(t_0)$ and simplifying finally yields

$$d_{1,\pm}(t_0) = \frac{\sqrt{2}\pi}{kK(k)} \sum_{m=-\infty}^{\infty} \frac{q^{m+\frac{1}{2}}}{1+q^{2m+1}} \frac{2\sqrt{2}\pi i\Lambda}{1-\Lambda^2}$$

$$\cdot \frac{(2m+1)^2\pi^2 h^{\frac{1}{4}}}{12K^2(k)}[1 + 2k^2\sqrt{h} - \sqrt{h}]e^{\frac{i(2m+1)\pi h^{\frac{1}{4}} t_0}{2K(k)}} . \tag{4.73}$$

But the bracketed expression can be simplified further by making use of Eq. (4.49), which gives

$$1 + 2k^2\sqrt{h} - \sqrt{h} = 1 + \sqrt{h} - 1 - \sqrt{h} = 0, \tag{4.74}$$

i.e., $d_{1,\pm}(t_0) \equiv 0$ for all h!

4.4 Notes and references

Armed with the insight of the previous chapter, the above discussion introduced a convenient tool for the detection of transversal intersections between stable and unstable manifolds when perturbing from an integrable system. In particular, the idea of measuring the separation of the manifolds originated in the work of Melnikov[1], and the resulting method is often referred to as the Melnikov method. Melnikov's work was then further developed by Holmes and Greenspan in a series of papers, see for example Greenspan[2] and Guckenheimer & Holmes[3]. As we shall see in later chapters, similar ideas appear in higher-dimensional calculations.

The method of solving the variational equations presented in this chapter has its origin in the work of Chow *et al*[4] where the Melnikov integral is derived using functional analytic conditions on the existence of a solution to the variational equation. We further note that many of the results in this chapter appear in a paper by the present author, see Dankowicz[5]. The same paper also shows that a similar analysis is possible for the detection of subharmonic periodic orbits inside the unperturbed homoclinic orbit. This is, of course, related to the subharmonic Melnikov method described in Guckenheimer & Holmes[3].

1. V. K. Melnikov, "On the Stability of the Center for Time-Periodic Perturbations," *Transactions of the Moscow Mathematical Society* **12** (1963), pp. 1-56.
2. B.D. Greenspan, *Bifurcations in Periodically Forced Oscillations: Subharmonic and Homoclinic Orbits* (Ph.D. thesis, Cornell University, 1981).
3. J. Guckenheimer and P. Holmes, *Nonlinear Oscillations, Dynamical Systems, and Bifurcations of Vector Fields* (Springer-Verlag, 1990).
4. S.-N. Chow, J. K. Hale and J. Mallet-Paret, "An Example of Bifurcation to Homoclinic Orbits," *Journal of Differential Equations* **37** (1980), pp. 351-373.
5. H. Dankowicz, "Looking for Chaos. An Extension and Alternative to Melnikov's Method," *International Journal of Bifurcation and Chaos* **6**(3) (1996), pp. 485-496.

Chapter 5
APPLICATION - RADIATION PRESSURE PROBLEMS IN CELESTIAL MECHANICS. PART I

Originating in thoughts of the distant and apparently imperturbable motion of the stars, humanity's philosophical considerations have never left the celestial spheres. The immensity of the universe away from our little rock has been an existential challenge of great proportion. Fascination with the large- and small-scale structure of space and its inhabitants has not decreased with our increasing experimental and theoretical knowledge. Instead, the marvel remains and awaits further exploration.

But the universe is fascinating even in the most humble investigations. The chaotic dynamics discussed so far in this book appear to have bearing on the motion of actual celestial objects, be they planets or dust grains. Mentioned in the first chapter, the long-term stability of the solar system, i.e., the persistence of the orbital dynamics we observe today, is believed to be addressed by the KAM results of Chapter 2. Assuming that the many-body problem afforded by the solar system is a small perturbation from the integrable collection of two-body problems consisting of each planet and the sun, KAM theory implies the existence of invariant quasiperiodic motion of the type observed in the solar system. However, as mentioned in Chapter 2, the invariant tori are interspersed with stochastic regions, and in the present high-degree-of-freedom situation, these are connected to allow large state-space excursions. Estimates of the time of drift away from a given quasiperiodic motion suggest that stability is at least guaranteed over comfortably long time scales.

Similarly, chaotic dynamics are thought to be responsible for the existence of gaps in the asteroid belt, or rather in the distribution of semi-major axes of the asteroid orbits. Here the responsible dynamical features clearly act on shorter time scales than the age of the solar system, as evidenced by their presence today.

In this chapter we study a particularly simple problem in celestial mechanics which exhibits many of the features discussed in previous chapters. Moreover, it has practical implications on the dynamics of grains orbiting asteroids, although the present study only begins to address such issues. In the first section we consider the various perturbations to two-body motion that solar system objects are subject to. Particular emphasis is given to the radiation pressure originating in the flux of solar photons. We furthermore discuss the distribution of matter in the solar system focusing on asteroids and small interplanetary and circumasteroidal dust particles.

Section 2 applies the method of the previous chapter to a mathematical model of circumasteroidal dust dynamics. As indicated there, the perturbation due to radiation pressure results in a separation of the stable and unstable manifolds of second order in the strength of the perturbation. The analytical findings are further illustrated through numerical simulations in Section 3.

5.1 Physical considerations

5.1.1 Common solar system perturbations

In Chapter 1 we discussed the integrable dynamics of two spherical bodies interacting solely through gravity. We found that their motion was restricted to a plane which, in turn, moved with constant velocity through space. In particular, bounded orbits were ellipses with one focus at the bodies' common center of mass.

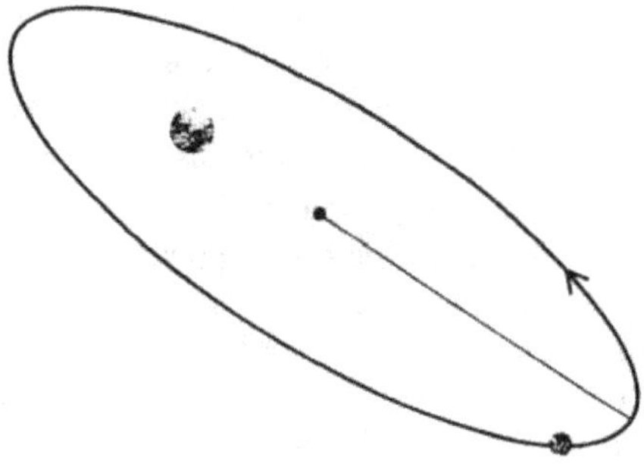

In the two-body problem orbits are Keplerian.

In general, celestial bodies do not simply interact with each other pairwise, but are instead influenced by many other bodies and fields in space. It is often the case that primarily one of the two bodies feels these additional forces, whereas the other is more or less unaffected. We refer to these situations as perturbed two-body problems. The perturbing forces can be of different origin.

Gravitational perturbations

We have shown that the gravitational interaction between spherical bodies follows the Newtonian law between particles. Nonsphericity of the primary, however, results in additional perturbations, so-called oblateness perturbations, to the motion of the satellite. One can show that these effects become negligible at large distances between the bodies.

In particular, let the origin of the coordinate system be positioned at the center of mass of the primary and denote its density and total mass by ρ and m_1,

The motion around a nonspherical primary is no longer Keplerian.

respectively. Its gravitational effects are then obtained from the potential

$$V(\boldsymbol{x}) = -G \int \frac{\rho(\boldsymbol{x}')}{|\boldsymbol{x} - \boldsymbol{x}'|} \mathrm{d}\boldsymbol{x}', \tag{5.1}$$

where the integration is over the entire volume of the primary. Furthermore, the force on a particle of mass m_2 at a point $\boldsymbol{x}$ is given by

$$\boldsymbol{F}_{grav} = -m_2 \frac{\partial V}{\partial \boldsymbol{x}}. \tag{5.2}$$

It is common to expand the integrand in Eq. (5.1) in terms of Legendre polynomials yielding:

$$V(\boldsymbol{x}) = -G \int \rho(\boldsymbol{x}') \sum_{n=0}^{\infty} P_n(\cos\theta) \left(\frac{|\boldsymbol{x}'|}{|\boldsymbol{x}|}\right)^n \mathrm{d}\boldsymbol{x}', \tag{5.3}$$

where θ is the angle between $\boldsymbol{x}$ and $\boldsymbol{x}'$.

Illustration

The first Legendre polynomial identically equals 1. Thus, to lowest order in $|\boldsymbol{x}|^{-1}$ we find

$$V(\boldsymbol{x}) = -\frac{Gm_1}{|\boldsymbol{x}|}, \tag{5.4}$$

which accords with our previous results. Furthermore, since

At large distances, the gravitational potential is essentially that of a particle at the center of the primary.

$$P_1(\cos\theta) = \cos\theta = \frac{\boldsymbol{x}\cdot\boldsymbol{x}'}{|\boldsymbol{x}|\,|\boldsymbol{x}'|}, \tag{5.5}$$

the term of order $|\boldsymbol{x}|^{-2}$ *vanishes identically. In fact,*

$$\frac{G}{|\boldsymbol{x}|^2}\boldsymbol{x}\cdot\int \rho(\boldsymbol{x}')\,\boldsymbol{x}'\mathrm{d}\boldsymbol{x}' = 0, \tag{5.6}$$

since the integral yields the position of the center of mass, which we took to be at the origin. Thus,

$$V(\boldsymbol{x}) = -\frac{Gm_1}{|\boldsymbol{x}|} + \mathcal{O}(|\boldsymbol{x}|^{-3}) \tag{5.7}$$

for the gravitational potential of a general body. •

The presence of additional bodies, so-called third-body interactions, are also often important. For example, as mentioned earlier, observed anomalies in the motion of Uranus lead to the prediction that yet another planet should exist outside the Uranus orbit. Similarly, much of the observed dynamics of the asteroids in the asteroid belt between Mars and Jupiter can be approached only if Jupiter's, and sometimes even Saturn's, gravitational influence is included.

The perturbations due to a third body can result in measurable deviations from elliptical motion.

Collisional perturbations

Objects in low-flying orbits around planets with atmosphere feel a drag due to their plowing through the atmosphere and to subsequent impacts with its molecules. The consequent slow-down can be rather substantial and rapidly lead to the decay of the orbits of these objects, which eventually collide with the planet and are destroyed. Another instance of collisional forces arises from the pressure of the solar wind, the flux of particles from the sun. This has been suggested as a potential source for propulsion.

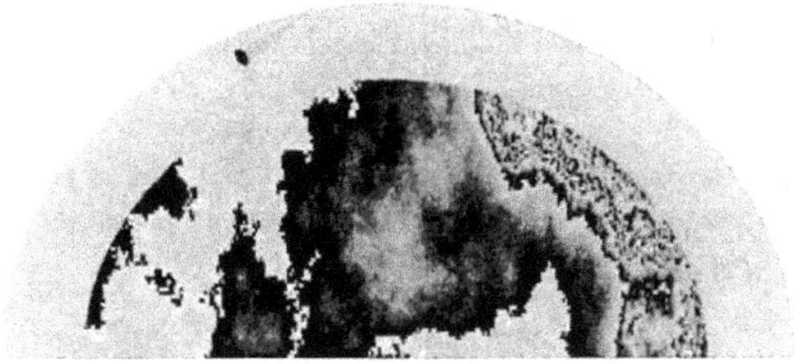

Particularly evident are the effects of the atmosphere on shooting stars.

Radiation forces

In addition to the flux of particles from the sun, a large amount of radiation is emitted in a wide spectrum of wavelengths. The absorption and subsequent re-emission of this radiative energy results in an effective perturbation of the motion of the absorbing object.

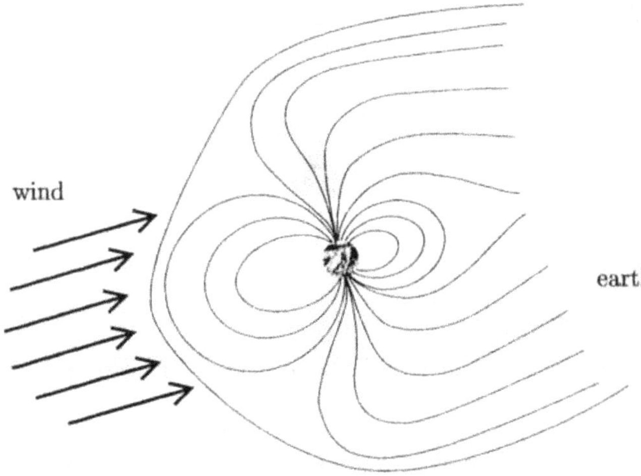

The solar wind carries with it the solar magnetic field which then collides with earth's magnetosphere to form a bow shock.

Illustration

We consider a particle of unit mass moving with velocity $\boldsymbol{v} = v\boldsymbol{n}$ in a reference frame in which the sun is at rest. An incoming beam of photons impinging on the particle will have the relativistic four-momentum

$$p = (E, E, 0, 0)^T \tag{5.8}$$

where E is the energy flux per unit time and unit area of the beam in geometrized units (i.e., normalized with respect to the speed of light). Because of the relativistic Doppler effect, we can write

$$E = E_{rest}(1 - \boldsymbol{e}_r \cdot \boldsymbol{v}) \tag{5.9}$$

where E_{rest} is the flux measured by a stationary observer and $\boldsymbol{e}_r$ is the radial unit vector. In the reference frame of the particle, the beam's four-momentum is given by

$$p' = \begin{pmatrix} \gamma & -v\gamma n_1 & -v\gamma n_2 & -v\gamma n_3 \\ -v\gamma n_1 & (\gamma-1)n_1^2+1 & (\gamma-1)n_1 n_2 & (\gamma-1)n_1 n_3 \\ -v\gamma n_2 & (\gamma-1)n_2 n_1 & (\gamma-1)n_2^2+1 & (\gamma-1)n_2 n_3 \\ -v\gamma n_3 & (\gamma-1)n_3 n_1 & (\gamma-1)n_3 n_2 & (\gamma-1)n_3^2+1 \end{pmatrix} p \tag{5.10}$$

where $\gamma = 1/\sqrt{1-v^2}$.

The energy transfer due to interactions with photons affects the dynamics of material particles.

When the particle interacts with the incoming radiation, only a fraction passes through unperturbed. The remaining radiation is either scattered according to some appropriate scattering law, or absorbed only to be re-emitted isotropically, since the particle is at rest in this frame. Clearly, the fraction that is absorbed has no net effect on the particle, whereas the scattered radiation does affect the particle, depending on its directional dependence. We account for this by assuming that a fraction Q of the original energy is lost in the interaction. In other words, the three-momentum of the outgoing beam is multiplied by $(1-Q)$. Inverting the above Lorentz transformation and forming the difference between the incoming and the

Radiation is scattered by the presence of the particle.

outgoing beam four-momentum, we obtain

$$\Delta \boldsymbol{p} = \frac{EQ}{1-v^2}\begin{pmatrix} vn_1 - v^2 \\ 1 - vn_1 + v^2(n_1^2 - 1) \\ v^2 n_1 n_2 - vn_2 \\ v^2 n_1 n_3 - vn_3 \end{pmatrix}. \tag{5.11}$$

This change in four-momentum of the beam per unit time and unit area is equal and opposite to the change in four-momentum of the particle per unit time and unit area. This, however, equals the four-vector pressure experienced by the particle. Thus, expanding the last three components of $\Delta \boldsymbol{p}$ *in* v *we obtain*

$$\boldsymbol{F}_{rad} = E_{rest} AQ \left[\boldsymbol{e}_r - \boldsymbol{v}\cdot\boldsymbol{e}_r - \boldsymbol{v} + \mathcal{O}(\boldsymbol{v}^2)\right], \tag{5.12}$$

where A *is the area of the particle. Finally, we return to normal units by dividing* E_{rest} *and* $\boldsymbol{v}$ *by* c*, the speed of light, and write*

$$\boldsymbol{F}_{rad} = \frac{E_{rest}AQ}{c}\left[(1 - 2\frac{\dot{r}}{c})\boldsymbol{e}_r - \frac{r\dot{\theta}}{c}\boldsymbol{e}_\theta\right] \tag{5.13}$$

plus higher order terms. The velocity-dependent term in this expression is known as the ***Poynting-Robertson drag****, and the remaining term as* ***radiation pressure****. •*

The energy flux per unit area and unit time decays as r^{2}, since

$$E_{rest} = \frac{L}{4\pi r^2} \tag{5.14}$$

where r is the distance from the sun and L is the solar luminosity. It is therefore convenient to express the radiation pressure as a fraction of the gravitational attraction to the sun at distance r. In fact, the ratio between radiation pressure and gravitation is

$$\nu = \frac{LAQ}{4\pi r^2 c}\frac{r^2}{V\rho GM} = \frac{L}{16\pi^2 GMc}\frac{Q}{R\rho} \approx 5.7 \times 10^{-7}\frac{Q}{R\rho}\mathrm{kg\,m}^{-2}, \tag{5.15}$$

where R and ρ are the radius and density, respectively, of the particle. The constant Q depends on the optical properties of the particle, as well as its size and the spectrum of the solar radiation. For particles larger than tens of microns, Q is essentially size- and constituent-independent and approximately equal to 1. For smaller particles, careful numerical integrations are necessary to obtain estimates of Q for different materials as a function of particle size. In the following sections we consider the effects of radiation pressure on particles of fractions-of-millimeter-size, and hence assume $Q \approx 1$.

Electromagnetic perturbations

Solar radiation impacting on small particles can lead to permanent ionization, thus creating charged particles. In interplanetary space, these small particles experience the effects of the heliosphere, the electromagnetic field associated with the sun. Similarly, planetary magnetic fields affect the dynamics of circumplanetary charged objects. For example, the earth's magnetic field results in high concentrations of electrons and ions in the so-called van Allen belts.

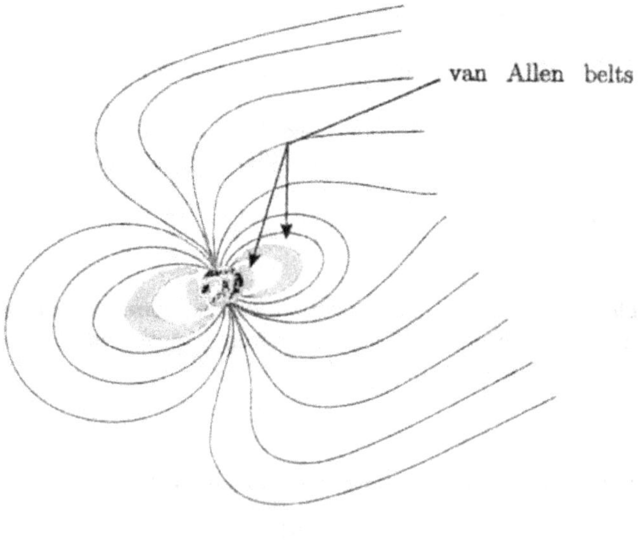

Charged particles bounce back and forth between the magnetic poles while spiralling through the van Allen belts.

5.1.2 The bodies of the solar system

In addition to the major planets and their moons, the solar system contains a plethora of smaller objects. These range from larger asteroids with typical sizes of hundreds of kilometers down to microscopic dust particles in circumplanetary orbits. Clearly, the perturbations discussed in the previous section will affect only some of these objects and their effects vary greatly.

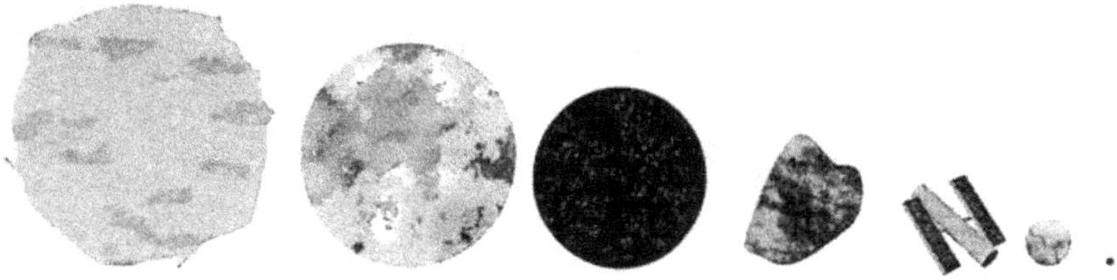

Some typical solar system members. Each step to the right is equivalent to a reduction of radius by a factor of a 100.

The asteroids are mainly collected in a region of the solar system between the orbits of Mars and Jupiter known as the asteroid belt. This is thought to be a remnant of the original solar nebula for which the accretion of planetesimals to form terrestrial planets was interrupted by the early appearance of Jupiter. Having swept up large amounts of material, Jupiter's gravitational influence further served to disturb the orbits of many of the planetesimals in its vicinity, thus causing their orbital eccentricities to increase to the point where no further accretion was possible. This accounts for the fact that the total asteroidal mass is only about 5 percent of the moon's. It is believed that the present asteroid belt has remained more or less unchanged since the formation of the solar system, albeit smaller collisions continue to occur.

Recently, the exploration of space has begun to include the asteroids. In particular, recent near-encounters by the Galileo satellite with the asteroids Gaspra and Ida and the upcoming approach of the NEAR satellite with Eros have brought asteroid science into the astronomical focus. Fictional and perhaps visionary accounts of future solar system development involve industrial mining of the asteroids. Their low mass and consequent small escape velocities (i.e., velocities required to completely escape the asteroid) might make such endeavors commercially viable.

In addition to the asteroid-size objects, smaller dust particles with sizes ranging from fractions of microns to several centimeters occupy interplanetary space. Their presence is evidenced by the large amount of dust that impinges on earth every year. Also, microscopic impact craters on glass spheres found on the moon, and the zodiacal light thought to be the result of scattered solar radiation are further indicators of such dust particles. These particles are generally thought to derive from cometary tails, but also possibly from collisions between larger objects, say in the asteroid belt.

The discussion of the previous section indicates that forces derived from the solar photon flux ought to be particularly important to the dynamics of interplanetary dust. Clearly, if $\nu > 1$ then radiation pressure will outweigh the gravitational attraction to the sun, with the resulting hyperbolic orbit blowing the dust out of the solar system. In fact, even if $\nu < 1$ it is possible that dust initially bound to comets on elliptic orbits will gain a sufficient impulse to end up on a hyperbolic orbit, even though the interaction with the sun is still attractive.

Heating and collisions with solar particles lead to the slow disintegration of circumstellar comets.

Illustration

In Chapter 1 we found that the semimajor axis, a, of a conical orbit was related to the total energy, E, through

$$E = -\frac{\mu}{2a}. \tag{5.16}$$

Clearly, particles on elliptical orbits have $E < 0$, whereas $E > 0$ implies unbounded motion. At the closest approach to the primary, the relative distance is

$$\varrho_0 = a(1-e) \tag{5.17}$$

and the radial momentum obviously zero. If at that point a particle is released from the comet and thus experiences a diminished attraction to the sun so that $\mu \to \mu' = \mu(1-\nu)$, then its total energy becomes

$$H = \frac{1}{2}\frac{l^2}{a^2(1-e)^2} - \frac{\mu'}{a(1-e)} = \frac{\mu}{a(1-e)}\left(\frac{e-1}{2} + \nu\right). \tag{5.18}$$

This quantity is positive provided

$$\nu > \frac{1-e}{2}. \tag{5.19}$$

Clearly, for comets on highly elongated orbits the right-hand side will be very small. Thus, even small values of ν could suffice to lead to the ejection of particles from the solar system. •

The Poynting-Robertsson drag also affects the dynamics of dust particles. For interplanetary dust it would tend to decrease the energy and angular momentum of a particle, thus eventually leading to annihilation in the sun. Since the drag force is of higher order in the ratio between velocity and the speed of light, it is expected to be important over longer times, say tens to hundreds of thousands of years. For shorter times, however, it initially can be ignored.

In the following section we consider dust particles orbiting asteroids and their dynamics as a consequence of the influence of radiation. The goal is to apply the theory of the previous chapter to a suitable mathematical model of the dust dynamics and show the possibility of highly complicated behavior of the particles.

5.2 The planar two-body problem with radiation pressure

5.2.1 The physical model

In this section we consider a simplified model of the planar motion of a satellite in orbit around a primary, under the additional influence of radiation pressure of solar origin. To first approximation, we neglect the gravitational influence of the distant sun. Clearly, if both primary and satellite experience identical accelerations, then these cancel out of the equations describing the relative motion. While the solar gravitation is not homogeneous in the neighborhood of the primary, this is a reasonable initial simplification, provided the distance to the sun is sufficiently large.

Moreover, the radiation pressure is taken to be constant in magnitude throughout the region of motion of the satellite but with direction rotating at constant rate. Again, at sufficient distance from the sun, the radiation field will be essentially homogeneous, but its direction will change with the motion of the primary around the sun. Similarly, magnitude variations in time due to the circumstellar motion can be neglected for sufficiently circular orbits. No other perturbations will be included; for example, nonsphericity of the primary is not considered. In the case of asteroids with highly irregular shapes this approximation might not be acceptable. Nevertheless, we are primarily interested in illustrating qualitative effects for which the present model will suffice.

In Chapter 7 a more complicated model, including some of the effects neglected here as well as an additional degree of freedom, will be addressed. As noted in previous chapters, higher-degree-of-freedom dynamics offer additional qualitative behavior not observed in two-degree-of-freedom motion. This is the focus of discussion in the last chapters of this book.

5.2.2 A mathematical formulation

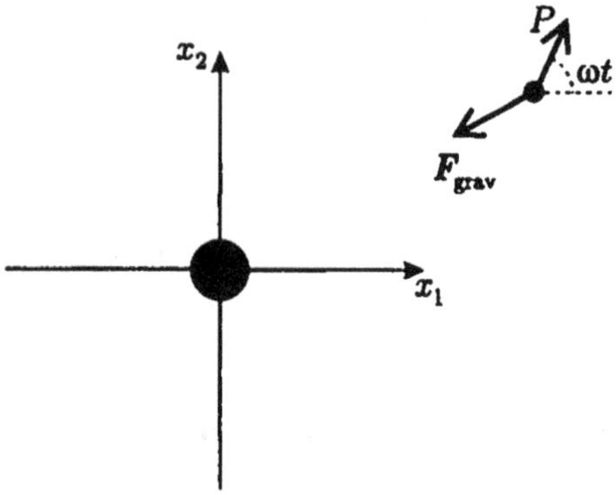

A periodically perturbed planar two-body problem.

Denote the magnitude of the acceleration due to radiation pressure by P and assume that its line of action rotates with constant rate ω. Then, Eq. (2.47) with $V = -Px_1 \cos\omega t - Px_2 \sin\omega t$ yields

$$H(\boldsymbol{x}, \boldsymbol{y}, t) = \frac{1}{2}|\boldsymbol{y}|^2 - \frac{\mu}{r} - Px_1 \cos\omega t - Px_2 \sin\omega t, \tag{5.20}$$

where $r = |\boldsymbol{x}| = \sqrt{x_1^2 + x_2^2}$. Here, $\boldsymbol{x}$ and $\boldsymbol{y}$ are conjugate coordinates and $\boldsymbol{x}$ is simply the Cartesian position vector. Eliminating the r^{-1} singularity is likely to make the subsequent analysis more tractable. We note also that the singularity removal is even more important in numerical studies of this Hamiltonian system. Following the theory in Section 2.2.3 we obtain

$$H_s(\boldsymbol{x}, x_0, \boldsymbol{y}, y_0) = \frac{r}{2}|\boldsymbol{y}|^2 - \mu - Pr(x_1 \cos\omega x_0 + x_2 \sin\omega x_0) + y_0 r \tag{5.21}$$

where $y_0 \equiv -H$, i.e., $H_s \equiv 0$.

The nature of the perturbation suggests a change of reference to a frame rotating with the radiation pressure so that its line of action remains constant. This is accomplished by the generating function

$$S(\bar{\boldsymbol{x}}, \bar{x}_0, \boldsymbol{y}, y_0) = -(y_0, y_1, y_2) \begin{pmatrix} \bar{x}_0 \\ \bar{x}_1 \cos\omega\bar{x}_0 - \bar{x}_2 \sin\omega\bar{x}_0 \\ \bar{x}_1 \sin\omega\bar{x}_0 + \bar{x}_2 \cos\omega\bar{x}_0 \end{pmatrix} \tag{5.22}$$

of the form (2.37). Substituting the corresponding coordinate transformations from Eqs. (2.38-2.39) yields the transformed Hamiltonian

$$\bar{H}_s(\bar{\boldsymbol{x}}, \bar{x}_0, \bar{\boldsymbol{y}}, \bar{y}_0) = \frac{r}{2}|\bar{\boldsymbol{y}}|^2 - \mu - P\bar{x}_1 r + \bar{y}_0 r + \omega r(\bar{y}_1\bar{x}_2 - \bar{y}_2\bar{x}_1), \tag{5.23}$$

where $r = |\bar{\boldsymbol{x}}|$.

The Hamiltonian can be further simplified using the canonical Levi-Civita transformation

$$S_{KS}(u_0, \boldsymbol{u}, \bar{y}_0, \bar{\boldsymbol{y}}) = -(\bar{y}_0, \bar{y}_1, \bar{y}_2)\begin{pmatrix} u_0 \\ u_1^2 - u_2^2 \\ 2u_1u_2 \end{pmatrix} \tag{5.24}$$

to the canonical variables (u, w). Substituting this into the Hamiltonian yields

$$H_{LC}(\boldsymbol{u}, u_0, \boldsymbol{w}, w_0) = \frac{1}{8}|\boldsymbol{w}|^2 - \mu + P(u_2^4 - u_1^4) + w_0 r + \frac{\omega}{2} r(u_2w_1 - u_1w_2) \tag{5.25}$$

where $r = |\boldsymbol{u}|^2$. Since, u_0 is cyclic, it follows that w_0 is invariant. We apply the rescaling

$$\boldsymbol{u} = \sqrt{\frac{e}{2P}}\bar{\boldsymbol{u}}, \ \boldsymbol{w} = \frac{e}{2\sqrt{P}}\bar{\boldsymbol{w}}, \ s = \sqrt{\frac{2}{e}}\bar{s}, \tag{5.26}$$

where $e = |w_0|$, which yields

$$\bar{H}_{LC}(\bar{\boldsymbol{u}}, \bar{\boldsymbol{w}}) = \frac{1}{8}|\bar{\boldsymbol{w}}|^2 + 2|\bar{\boldsymbol{u}}|^2 \text{sign}(w_0) + (\bar{u}_2^4 - \bar{u}_1^4) + \varepsilon|\bar{\boldsymbol{u}}|^2(\bar{w}_1\bar{u}_2 - \bar{w}_2\bar{u}_1), \tag{5.27}$$

where

$$\bar{H}_{LC} \equiv \frac{4P\mu}{e^2}, \text{ and } \varepsilon = \frac{\sqrt{e}}{2\sqrt{2P}}\omega. \tag{5.28}$$

To simplify notation, the bar will be omitted in the discussion to follow and, furthermore, we write $H_\varepsilon = \bar{H}_{LC}$. The sequence of transformations has left us with a two-degree-of-freedom Hamiltonian describing the motion on the constant w_0 manifold for which H_ε is given above. Reversing the sequence of coordinate transformations we obtain

$$w_0 = \omega(x_1y_2 - x_2y_1) - H \equiv \text{constant}, \tag{5.29}$$

usually referred to as the Jacobi integral.

5.2.3 The unperturbed flow

When $\omega = 0$, i.e., $\varepsilon = 0$, the Hamiltonian separates into

$$H_0(\boldsymbol{u}, \boldsymbol{w}) = F(u_1, w_1) + G(u_2, w_2), \tag{5.30}$$

where

$$F(u_1, w_1) = \frac{1}{8}w_1^2 + 2u_1^2 \text{sign}(w_0) - u_1^4 \tag{5.31}$$

and

$$G(u_2, w_2) = \frac{1}{8}w_2^2 + 2u_2^2\text{sign}(w_0) + u_2^4, \tag{5.32}$$

and is clearly integrable for all values of e, P, and μ.

From the equation of motion for w_1 it is easy to see that $\text{sign}(w_0) = -1$ implies the absence of bounded motion, and thus the particle is quickly blown away from the primary. Since we are interested in orbits that remain in the vicinity of the primary for large times, we henceforth restrict attention to $\text{sign}(w_0) = 1$. It is clear that bounded orbits correspond to motion along closed orbits in both degrees-of-freedom. Since the periods of such motion differ in the two degrees-of-freedom, the combined motion is generally quasiperiodic. We note that the (u_1, w_1)-portrait is identical to that studied in the previous chapter.

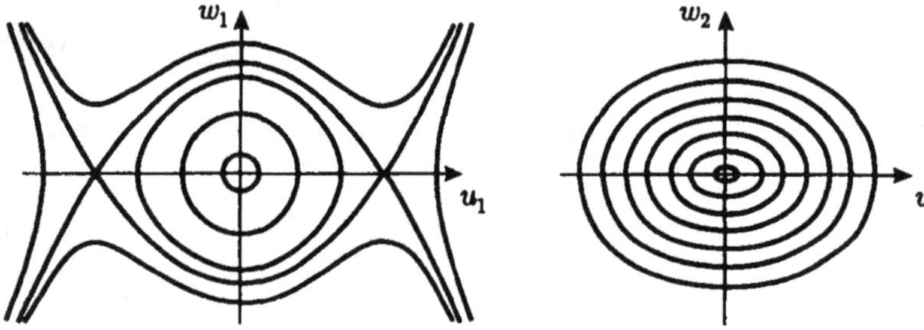

The motion in the first degree-of-freedom contains a heteroclinic cycle when $w_0 > 0$. The motion in the second degree-of-freedom is always bounded.

Illustration

To transform to action-angle coordinates in the (u_2, w_2)-degree-of-freedom, we consider Hamilton-Jacobi's equation

$$\frac{1}{8}\left(\frac{\partial S}{\partial u_2}(I, u_2)\right)^2 + 2u_2^2 + u_2^4 = \Psi(I). \tag{5.33}$$

In particular, we restrict attention to a particular closed orbit for which $u_2 \in [-a, a]$. Solving this we obtain

$$S(I, u_2) = 2\sqrt{2}\int_{-a}^{u_2} \sqrt{\Psi(I) - 2s^2 - s^4}ds, \tag{5.34}$$

where we have chosen the positive root corresponding to the upper half of the closed orbit. Eq. (2.31) then gives

$$\varphi = \frac{\partial S}{\partial I}(I, u_2) = \sqrt{2}\Psi'(I)\int_{-a}^{u_2} \frac{1}{\sqrt{\Psi(I) - 2s^2 - s^4}}ds, \tag{5.35}$$

where we used the fact that the square root vanishes for $s = -a$. Since $\varphi \in \mathbb{S}^1$, it is convenient to set $\varphi(u_2 = a) = \pi$. Thus,

$$\pi = \sqrt{2}\Psi'(I)\int_{-a}^{a}\frac{1}{\sqrt{\Psi(I)-2s^2-s^4}}\mathrm{d}s, \tag{5.36}$$

which we can integrate to yield

$$\pi I + C = 2\sqrt{2}\int_{-a}^{a}\sqrt{\Psi(I)-2s^2-s^4}\mathrm{d}s, \tag{5.37}$$

where C is an integration constant. The integral on the right-hand side of this expression actually equals half the area inside the closed orbit. If we let $I \to 0$ as the area shrinks to zero, we find $C = 0$, and

$$I = \frac{2\sqrt{2}}{\pi}\int_{-a}^{a}\sqrt{\Psi(I)-2s^2-s^4}\mathrm{d}s. \tag{5.38}$$

The integrals in Eq. (5.35, 5.36) can actually be given in closed form. For simplicity we write

$$\Psi(I) - 2s^2 - s^4 = (a^2 - s^2)(b^2 + s^2), \tag{5.39}$$

where

$$a^2 = \sqrt{1+\Psi(I)} - 1, \text{ and } b^2 = \sqrt{1+\Psi(I)} + 1. \tag{5.40}$$

Then, the substitution

$$t = \frac{ab}{\sqrt{a^2+b^2}}\frac{\mathrm{sn}[x,k]}{\mathrm{dn}[x,k]},\ k^2 = \frac{a^2}{a^2+b^2} \tag{5.41}$$

and intricate but straightforward algebra yield

$$u_2 = -a\mathrm{cn}\left[\frac{\varphi\sqrt{a^2+b^2}}{\sqrt{2}\Psi'(I)}, k\right]. \tag{5.42}$$

Since cn has period $4K(k)$ and $\varphi \in [0, 2\pi]$ it follows that

$$\Psi'(I) = \frac{\pi(1+\Psi(I))^{\frac{1}{4}}}{2K(k)}. \tag{5.43}$$

Eliminating a and b we finally obtain

$$u_2 = -\sqrt{2}k(1+\Psi(I))^{\frac{1}{4}}\mathrm{cn}\left[\frac{2K(k)\varphi}{\pi}, k\right], \tag{5.44}$$

where

$$k^2 = \frac{\sqrt{1+\Psi(I)}-1}{2\sqrt{1+\Psi(I)}}. \tag{5.45}$$

Solving Eq. (5.33) for $\partial S/\partial u_2$ yields

$$w_2 = \frac{\partial S}{\partial u_2} = 4\sqrt{2}k\sqrt{1+\Psi(I)}\mathrm{sn}\left[\frac{2K(k)\varphi}{\pi}, k\right] \mathrm{dn}\left[\frac{2K(k)\varphi}{\pi}, k\right], \tag{5.46}$$

completing the derivation of the canonical transformation. •

We note that the frequency of the periodic motion with Hamiltonian $G = g$ in the (u_2, w_2)-system is

$$\omega_2(g) = \frac{\pi(1+g)^{\frac{1}{4}}}{2K(k)}, \tag{5.47}$$

where

$$k^2 = \frac{\sqrt{1+g}-1}{2\sqrt{1+g}}. \tag{5.48}$$

Similarly, we find the frequency

$$\omega_1(f) = \frac{\pi(1+\sqrt{1-f})}{2\sqrt{2}K(\tilde{k})} \tag{5.49}$$

where

$$\tilde{k} = \frac{1-\sqrt{1-f}}{1+\sqrt{1-f}} \tag{5.50}$$

and $F = f$ on the periodic motion. Clearly, $\omega_1 \to 0$ as $f \to 1$, which corresponds to approaching the heteroclinic motion along the separatrix. It is straightforward to show that $\omega_1'(f), \omega_2'(g) \neq 0$ for $f, g > 0$. Thus,

$$\det\left(\frac{\partial \boldsymbol{\omega}}{\partial \boldsymbol{I}}\right) = \det\left(\frac{\partial^2 H_0}{\partial \boldsymbol{I}^2}\right) \neq 0, \tag{5.51}$$

where $\boldsymbol{I}$ denotes the action variables for the unperturbed Hamiltonian H_0. The discussion from Chapter 2 now applies and we conclude that a majority of unperturbed nonresonant tori, corresponding to noncommensurable periodic motion in the two degrees-of-freedom, survive any small perturbations. It further follows that the origin remains stable, since it is enclosed within arbitrary small invariant tori, which prevent orbits from escaping the vicinity of the origin.

The fate of resonant tori depends more sensitively on the perturbation. If the perturbed system is integrable, they will clearly all persist. In general, this is not the case, and a more careful analysis shows the persistence of stable and unstable lower-dimensional tori in the spaces between the nonresonant tori. The unstable lower-dimensional tori have stable and unstable manifolds that intersect transversely and the resulting chaotic dynamics manifests itself as the stochastic layers introduced in Chapter 2.

Naturally, the dynamics of the perturbed system are likely to be dependent on the fate of the homoclinic manifold in the (u_1, w_1)-system. We shall shortly apply the methods of the previous chapter, but first we reexpress the two-degree-of-freedom system in a more suitable form.

5.2.4 Reduction of order

The unperturbed Hamiltonian can now be written

$$H_0(z, I) = F(z) + \Psi(I) \tag{5.52}$$

where $z = (u_1, w_1)$. Consequently,

$$H_\varepsilon(z, I, \varphi) = H_0(z, I) + \varepsilon H_1(z, I, \varphi). \tag{5.53}$$

In the previous chapter we showed how a perturbed two-degree-of-freedom Hamiltonian system of the kind in Eq. (5.53) can be transformed to a periodically forced single-degree-of-freedom system with Hamiltonian $L_\varepsilon = \sum_k \varepsilon^k L_k$, where in this case

$$L_0(z; h) = -\Psi^{-1}(h - F(z)) \tag{5.54}$$

$$L_1(z, \varphi; h) = \frac{H_1(z, \varphi, -L_0(z; h))}{\omega_2(-L_0(z; h))} \tag{5.55}$$

and

$$L_2(z, \varphi; h) = -\frac{1}{2\omega_2(-L_0(z; h))}\frac{\partial}{\partial I}\left(\frac{(H_1(z, \varphi, I))^2}{\omega_2(I)}\right)\Bigg|_{I=-L_0(z;h)}, \tag{5.56}$$

where $\omega_2(I) = \Psi'(I)$, and $h = H_\varepsilon$ on a particular energy manifold.

From Eq. (5.54) it follows that

$$\frac{\partial L_0}{\partial z} = \frac{1}{\omega_2(-L_0(z; h))}\frac{\partial F}{\partial z}. \tag{5.57}$$

For a given unperturbed motion in the (u_1, w_1)-system, F is conserved, and hence, L_0 is also conserved. Thus, the Hamiltonian system corresponding to L_0 has identical dynamics to those of F bar for the rescaling of time through the ω_2 factor. Consequently, the unperturbed phase portrait for the Hamiltonian L_ε is that shown previously. In particular, there exist two heteroclinic manifolds connecting the two saddle points to form a heteroclinic cycle. We expect this structure to break up as a consequence of the perturbation.

In the previous chapter, we derived appropriate measures for the splitting of the manifolds. As in the example problem described there, it follows from Eq. (5.27) that H_ε is invariant under the reflection $(u_1, \varphi) \to (-u_1, -\varphi)$. This, in turn, implies that L_ε is preserved under the same transformation; i.e., the hyperbolic periodic orbits near the original saddle points are again mirror images, and the homoclinic theory applies.

5.2.5 *First-order separation*

We consider the effect of perturbations on the upper heteroclinic connection. $F = 1$ implies that $I_0 = -L_0(\boldsymbol{z}_0; h) = \Psi^{-1}(h-1)$, and consequently

$$\Omega_0 = \omega_2(I_0) = \frac{\pi h^{\frac{1}{4}}}{2K(k)} \tag{5.58}$$

where

$$k^2 = \frac{\sqrt{h}-1}{2\sqrt{h}} \tag{5.59}$$

cf. Eq. (4.49). From Eq. (4.50) we find

$$\boldsymbol{z}_0^T(\varphi - \varphi_0) = \left(\tanh\left(\frac{\varphi - \varphi_0}{\sqrt{2}\Omega_0}\right), 2\sqrt{2}\mathrm{sech}^2\left(\frac{\varphi-\varphi_0}{\sqrt{2}\Omega_0}\right)\right) \tag{5.60}$$

where again, $\boldsymbol{\xi} = (0, 2\sqrt{2})$ is the point of the unperturbed manifold on which our perturbation analysis will be based.

Eq. (5.55) implies that

$$\frac{\partial L_1}{\partial \boldsymbol{z}} = \frac{1}{\omega_2(I)}\left[\frac{\partial H_1}{\partial \boldsymbol{z}}(\boldsymbol{z},\varphi,I) - \frac{\partial H_1}{\partial I}(\boldsymbol{z},\varphi,I)\frac{\partial L_0}{\partial \boldsymbol{z}}(\boldsymbol{z};h) + \frac{\omega_2'(I)H_1(\boldsymbol{z},\varphi,I)}{\omega_2(I)}\frac{\partial L_0}{\partial \boldsymbol{z}}(\boldsymbol{z};h)\right] \tag{5.61}$$

where $I = -L_0(\boldsymbol{z};h)$. The Poisson-bracket $\{L_0, L_1\}_{\boldsymbol{z}}$ is now easily evaluated and we obtain

$$\int_{-\infty}^{\infty}\{L_0,L_1\}(\boldsymbol{z}_0(\varphi-\varphi_0),\varphi;h)\mathrm{d}\varphi = \frac{1}{\Omega_0^2}\int_{-\infty}^{\infty}\{F,H_1\}(\boldsymbol{z}_0(\varphi-\varphi_0),\varphi,I_0)\mathrm{d}\varphi. \tag{5.62}$$

Substitution of the expressions for F and H_1, and partial integration, eventually yields

$$d_1(\varphi_0) = \frac{2\sqrt{2}}{\Omega_0^2}\int_{-\infty}^{\infty}\mathrm{sech}^2\left(\frac{\varphi-\varphi_0}{\sqrt{2}\Omega_0}\right)w_2(\varphi,I_0)\left[\tanh^2\left(\frac{\varphi-\varphi_0}{\sqrt{2}\Omega_0}\right) + u_2^2(\varphi,I_0)\right]\mathrm{d}\varphi \tag{5.63}$$

where u_2 and w_2 are given in Eqs. (5.44-5.46). Referring back to Eq. (4.51), we see that, as promised, the first-order separation of the manifolds is indeed identical to that in the example of Chapter 4, apart from a constant factor. In particular, this integral was found to vanish identically in φ_0 and h. Thus, we again find that the perturbed manifolds coincide to first order and it becomes necessary to go to higher order to discern the presence of transversal intersections between the manifolds.

5.2.6 Second-order calculations

As before, it becomes necessary to solve the first variational equation, since evaluating $\tilde{d}_2$ requires knowledge of z_1. We immediately find

$$(\phi_{11}(\varphi-\varphi_0),\phi_{12}(\varphi-\varphi_0))$$
$$=\frac{1}{\Omega_0}\left(\frac{1}{\sqrt{2}}\operatorname{sech}^2\left(\frac{\varphi-\varphi_0}{\sqrt{2}\Omega_0}\right),-4\tanh\left(\frac{\varphi-\varphi_0}{\sqrt{2}\Omega_0}\right)\operatorname{sech}^2\left(\frac{\varphi-\varphi_0}{\sqrt{2}\Omega_0}\right)\right). \tag{5.64}$$

Similarly,

$$\phi_{21}(\varphi-\varphi_0)=\frac{\varphi-\varphi_0}{\sqrt{2}}\left[\frac{\Omega_0'}{\Omega_0^2}+\frac{3}{16}\right]\operatorname{sech}^2\left(\frac{\varphi-\varphi_0}{\sqrt{2}\Omega_0}\right)$$
$$+\frac{\Omega_0}{16}\sinh\left(\sqrt{2}\frac{\varphi-\varphi_0}{\Omega_0}\right)+\frac{3\Omega_0}{16}\tanh\left(\frac{\varphi-\varphi_0}{\sqrt{2}\Omega_0}\right) \tag{5.65}$$

and

$$\phi_{22}(\varphi-\varphi_0)=-4(\varphi-\varphi_0)\left[\frac{\Omega_0'}{\Omega_0^2}+\frac{3}{16}\right]\tanh\left(\frac{\varphi-\varphi_0}{\sqrt{2}\Omega_0}\right)\operatorname{sech}^2\left(\frac{\varphi-\varphi_0}{\sqrt{2}\Omega_0}\right)$$
$$+\frac{\sqrt{2}\Omega_0}{4}\cosh\left(\sqrt{2}\frac{\varphi-\varphi_0}{\Omega_0}\right)+\frac{3\sqrt{2}\Omega_0}{4}\operatorname{sech}^2\left(\frac{\varphi-\varphi_0}{\sqrt{2}\Omega_0}\right), \tag{5.66}$$

where

$$\Omega_0'=\frac{d^2\Psi}{dI^2}(I_0)=\frac{\pi^2}{16K^2(k)}\left[\frac{1}{\sqrt{h}}-2\frac{E(k)-(1-k^2)K(k)}{K(k)(h-1)}\right]. \tag{5.67}$$

Since, $\phi_{12}(0)=\phi_{21}(0)=0$, it follows that $\alpha_1^{s,u}(\varphi_0)\equiv 0$. Furthermore, we easily obtain

$$\beta_1^{s,u}=\int_{\varphi_0}^{\infty}\frac{\partial L_1}{\partial z}(z_0(\varphi-\varphi_0),\varphi;h)\cdot(\phi_{11}(\varphi-\varphi_0),\phi_{12}(\varphi-\varphi_0))d\varphi$$
$$=\frac{1}{\Omega_0^2}\int_{\varphi_0}^{\infty}\{H_1,F\}(z_0(\varphi-\varphi_0),\varphi,I_0)d\varphi. \tag{5.68}$$

While the algebra is significantly complicated, it is possible to show, much as in the previous chapter, that for $\varphi_0=0$, the integrand in the evaluation of $\tilde{d}_2$ is odd in φ_0 and thus, $\tilde{d}_2(0)$ vanishes. It remains to show that this zero is simple. Again, we are unable to resolve this analytically, but numerical integrations show that the derivative with respect to φ_0 at $\varphi_0=0$ is indeed nonzero at least for particular values of h. It follows that the stable and unstable manifolds intersect transversely and, hence, that the dynamics on energy manifolds contain chaotic behavior as discussed in Chapter 3.

However, since the manifold splitting is of second order in ε, it is likely that its effects will only be significant over long time scales. The persistence of nonresonant invariant tori further limits the effects of the splitting to a narrow region near the original manifold. In the next section we show the results of some numerical investigations of the perturbed system, with particular focus on traces of chaotic behavior suggested by the analysis.

5.3 Some numerical results

In Chapter 1 we introduced a convenient tool for visualizing the flow of a two-degree-of-freedom Hamiltonian system. In particular, restricted to an energy submanifold, we considered plotting intersections of orbits with the surface $u_2 = 0$, $w_2 > 0$. In light of the reduction of order made use of in the present chapter, the corresponding map, taking one point of intersection to the subsequent, is essentially the period map of the reduced system. For vanishing perturbation, such a plot would thus be identical to the phase portrait shown above. Under small perturbations we have shown the persistence of the saddle points, now corresponding to non-trivial periodic orbits, and the splitting of their respective stable and unstable manifolds.

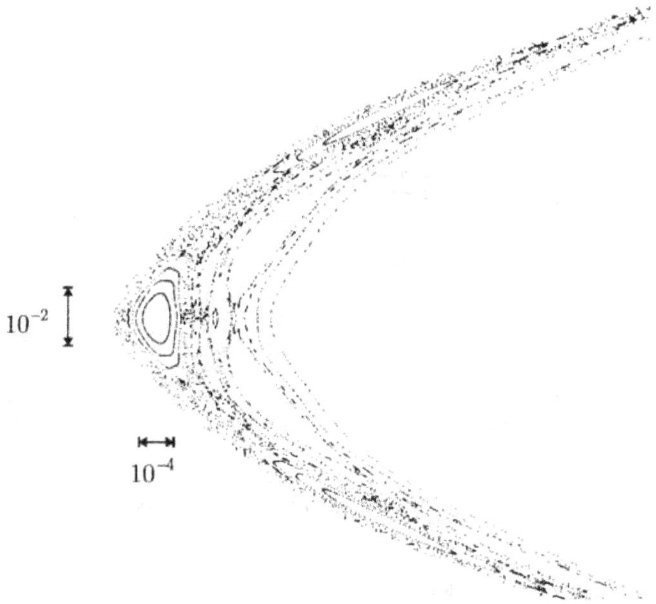

Intersections of different orbits on an energy submanifold with a suitable surface. Here, $\varepsilon = 0.05$.

As pointed out above, the persistent non-resonant tori inside the original heteroclinic cycle will constrain the chaotic region near the perturbed saddle orbits to a narrow region close to the original heteroclinic manifolds. In particular, we expect this to be most vividly clear in the vicinity of the perturbed saddle orbits. In the plot above, a number of initial conditions near the unperturbed saddle point have been integrated and the corresponding intersections plotted. We note the coexistence of chaotic bands and regular quasiperiodic motion in the surviving islands of closed curves. Of particular interest is the size of the regular islands. For a generic perturbation these are expected to be of $\mathcal{O}\left(\sqrt{\varepsilon}\right)$. In the present situation, on the other hand, their width appears to be at least one order of magnitude smaller. This is further evidence of the vanishing first-order splitting of the manifolds.

Another test for the presence of deterministic stochasticity is the existence of orbits with at least one positive Liapunov exponent. In particular, we consider an orbit in what appears to be a stochastic band in the plot above. In the graph below, the upper curve depicts the convergence of the sum

$$\frac{1}{n\Delta t}\sum_{k=1}^{n}\ln\delta_k\boldsymbol{x}_0 \tag{5.69}$$

as $n \to \infty$, where $\delta_k\boldsymbol{x}_0$ is obtained from the linearized flow as explained in Chapter 3. The non-negative limit supports the conclusions of the previous sections regarding the presence of chaotic motion and sensitive dependence on initial conditions. Its relatively small value is yet another indication that the chaotic effects are observable over longer time scales than generically expected. For comparison we include the corresponding evaluation about an orbit in the region of regular motion. Clearly, as predicted by the discussion in Chapter 3, the sum does not appear to converge to a positive value.

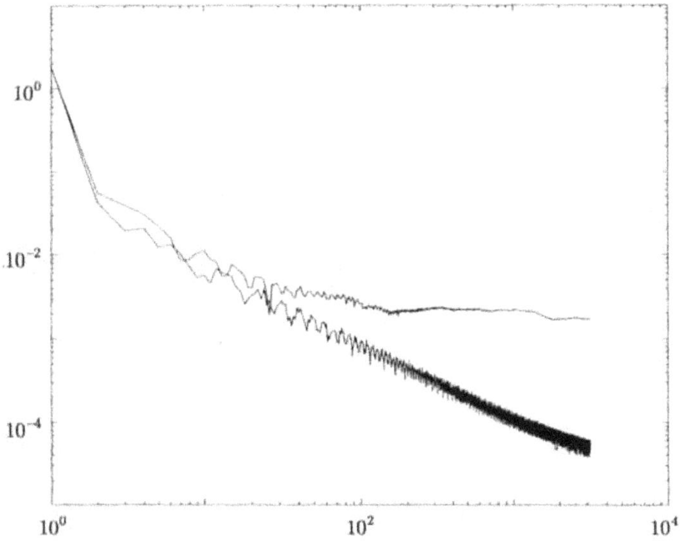

The largest Liapunov exponent of an orbit in the stochastic region converges to a positive value, indicating the sensitive dependence on initial conditions in that region.

5.4 Notes and references

In this chapter we introduced the main features of the solar system, its members and their dynamics. In particular, a variety of interactions between bodies and fields were considered as sources of perturbations to the integrable two-body problem. The emphasis on radiation pressure forces and their origins was motivated by the application discussed in the latter half of the chapter. A more detailed description of the constituents of the solar system and their interactions can be found in Taylor[1], with primary focus on the origin of the solar system. A similar discussion pertaining to the minor objects of the solar system is given in Burns[2]. Also, the derivation of the forces resulting from the flux of solar photons is based on work by Burns *et al*[3]. Finally, the discussion on the planar two-body problem considered in this chapter originated in work by the present author[4].

1. S. R. Taylor, *Solar System Evolution. A New Perspective* (Lunar and Planetary Institute, Cambridge University Press, 1992).
2. J. A. Burns ed., *Planetary Satellites* (The University of Arizona Press, 1977).
3. J. A. Burns, P. L. Lamy and S. Soter, "Radiation Forces on Small Particles in the Solar System," *ICARUS* **40** (1979), pp. 1-48.
4. H. Dankowicz, "The Two-Body Problem with Radiation Pressure in a Rotating Reference Frame," *Celestial Mechanics and Dynamical Astronomy* **61** (1995), pp. 287-313.

Chapter 6
GEOMETRY AND DYNAMICS IN MANY DEGREES OF FREEDOM

Complexity in social interactions increases dramatically with the number of involved parties; bear witness the saying "three's a crowd". Similarly, in the world of physics we have seen how an increase in the number of degrees-of-freedom in the gravitational n-body problem makes detailed knowledge of its dynamics practically impossible. In the other extreme of innumerable numbers of particles such as gases, the statistical methods of thermodynamics suggest large-scale order even where such is nowhere to be found in very small pockets of the gas. The desire to ascribe similar characteristics to large numbers of humans, be it nations or geographical regions lies at the heart of economics–the psychology of the mob.

In previous chapters we have shown how apparently random dynamics can arise in deterministic systems of very low dimension. In particular, we argued that the presence of transversal intersections of the stable and unstable manifolds of a periodic orbit were sufficient to imply the characteristics of chaos, e.g. sensitive dependence on initial conditions and periodic orbits of all periods. We applied the detection method derived in Chapter 4 to a particular physical example of astronomical (pardon the pun) importance.

The maxim in the first sentence of this section is further illustrated in this chapter through the examination of some unexpected properties of high-degree-of-freedom systems. Already the KAM results from Chapter 2 indicated the presence of highly stochastic motions interspersed between regular dynamics on the surviving tori when perturbing a system with global action-angle variables. It should come as no surprise that the available complexity is highly inflated if the unperturbed system contains inherent hyperbolicity, as did the presence of an unperturbed hyperbolic periodic orbit in the previous chapter. It is with this situation and the geometry of the perturbed stable and unstable manifolds of corresponding state-space features that this chapter is concerned.

In Section 1 we introduce the setting for this chapter. Armed with the appropriate structural assumptions, Sections 2 and 3 outline a proof of a persistence result in the presence of hyperbolicity, in the spirit of KAM. The consequent existence of stable and unstable manifolds of invariant tori is further studied in Section 4 where methods similar to those in Chapter 4 are applied to obtain perturbation expansions of solutions on these manifolds. The resulting expressions are used in Section 5 to investigate the presence of transversal intersections of these manifolds. Finally, Section 6 illustrates the possible consequences of such transversal intersections on the existence of large state-space excursions that persist to $\mathcal{O}(1)$ as the perturbation decreases in magnitude, only to vanish in the integrable limit.

6.1 General setup

In the previous chapter a particular two-degree-of-freedom Hamiltonian system was studied by reducing it to a periodically perturbed, nonautonomous one-degree-of-freedom Hamiltonian system to which the methods and results of the previous chapters applied. Although the analysis proved straightforward, it seems reasonable that similar conclusions could have been made by studying the original two-degree-of-freedom system, omitting the auxiliary step of reduction. This viewpoint will be taken in the present chapter. In fact, we shall describe results for arbitrary degrees-of-freedom provided the Hamiltonian flows decouple in an appropriate way.

In particular, consider the $(n+1)$-degree-of-freedom Hamiltonian $H(\boldsymbol{\varphi}, \boldsymbol{I}, \boldsymbol{z}; \varepsilon)$, where $\boldsymbol{z} = (q,p) \in \mathbb{R}^2$, $\boldsymbol{\varphi} \in \mathbb{T}^n$, $\boldsymbol{I} \in \mathbb{R}^n$, and H is real analytic in $\boldsymbol{\varphi}, \boldsymbol{I}, q, p$, and ε. Our first structural assumption is that, when $\varepsilon = 0$, the angles $\boldsymbol{\varphi}$ are cyclic, i.e., the unperturbed Hamiltonian $H_0(\cdot) = H(\cdot; 0)$ does not depend on $\boldsymbol{\varphi}$. Consequently, the action coordinates are invariants for the unperturbed system. Furthermore, we assume that for every $\boldsymbol{I}_0$ in some open set $\mathcal{U}$, the dynamics in the remaining degree-of-freedom possess a hyperbolic equilibrium that we, without loss of generality, assume to be at the origin. The unperturbed system thus contains one-dimensional stable and unstable manifolds in the $\boldsymbol{z}$ degree-of-freedom.

On the submanifold $\boldsymbol{z} = 0$ the motion is toral.

For $\boldsymbol{z} = \boldsymbol{0}$, the full flow is given by

$$\dot{\boldsymbol{\varphi}} = \boldsymbol{D_I} H_0(\boldsymbol{I}_0, 0) = \boldsymbol{\omega}, \tag{6.1}$$

i.e., the motion is on n-dimensional tori, $\tau_0(\boldsymbol{I}_0)$, parameterized by $\boldsymbol{I}_0$. The stable and unstable manifolds in the $\boldsymbol{z}$ degree-of-freedom now generalize to $(n+1)$-dimensional invariant manifolds, $\mathcal{W}^{s,u}(\tau_0(\boldsymbol{I}_0))$ on which orbits exponentially approach the torus as $t \to \infty$ and $-\infty$, respectively. Clearly, a homoclinic connection in the $\boldsymbol{z}$ degree-of-freedom corresponds to an $(n+1)$-dimensional homoclinic manifold in the energy manifold.

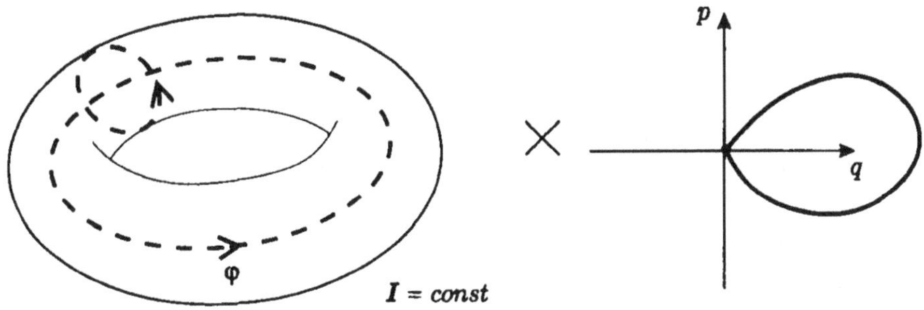

The full unperturbed dynamics decouple into tori cross motion in a single degree of freedom with a homoclinic orbit.

In other words, the unperturbed system is integrable since a complete set of pairwise Poisson-commuting invariants exists; e.g., the action variables $\boldsymbol{I}$ and a first integral in the $\boldsymbol{z}$ degree-of-freedom which, since it is a one-degree-of-freedom system, always exists. Contrary, however, to the situation in Chapter 2, where the unperturbed problem also was integrable, there does not exist a set of global action-angle variables, since the presence of the hyperbolic equilibrium prevents such to be constructed in the $\boldsymbol{z}$ degree-of-freedom. The energy manifolds are thus not foliated by tori, as in Chapter 2, but by the product of tori of dimension n and the energy manifolds in the $\boldsymbol{z}$ degree-of-freedom.

Illustration

Consider the Hamiltonian $H(\varphi, \boldsymbol{I}, \boldsymbol{z}; \varepsilon) = \boldsymbol{\omega}\cdot\boldsymbol{I} + p^2 - q^2 + \varepsilon H_1(\varphi, \boldsymbol{I}, \boldsymbol{z}; \varepsilon)$. *Clearly, the unperturbed Hamiltonian can be viewed as the product of n simple oscillators and a linear system with a saddle equilibrium at the origin.* •

The similarities between the tori and the hyperbolic periodic orbits of the previous chapter suggest that we focus our attention on the persistence of the tori and their manifolds and, in particular, on the existence of transversal intersections between these manifolds.

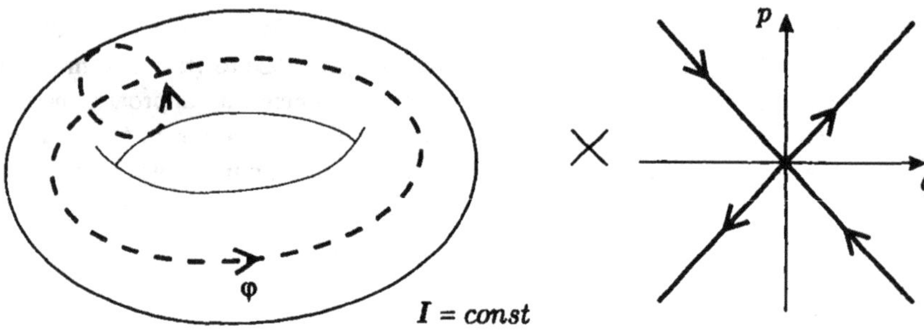

The unperturbed motion in the illustration decouples into tori cross linear dynamics.

6.2 A rapid convergence scheme

In Chapter 2 the persistence of the invariant tori under small perturbations was studied through the convergence of series solutions to Hamilton-Jacobi's equation. In this chapter we describe an alternative approach, more similar in spirit to the standard proof of the KAM theorem based on rapidly convergent approximation schemes, like the Newton-Raphson algorithm in numerical analysis. The method is most naturally presented in a number of separate steps. In particular, we begin by restricting attention to perturbations of the Hamiltonian

$$H_0(\boldsymbol{I}, \boldsymbol{z}) = h_0(\boldsymbol{I}) + \lambda_0 qp \tag{6.2}$$

where λ_0 is real and $\neq 0$ and $\boldsymbol{D}_I^2 h_0$ is invertible in a suitable region in state space. After presenting persistence results for this simplified Hamiltonian, we will argue that these naturally carry over to the general case through the use of appropriate canonical transformations.

6.2.1 *Topology*

Consider the family of complex cylinders

$$C_n = \{(\varphi, \boldsymbol{I}, q, p) \mid |\Im(\varphi)| \le \rho_n, |\boldsymbol{I} - \boldsymbol{I}_n^*| \le 4\varrho_n, |q|, |p| \le 6\varrho_n\}, n \in \mathbb{N}, \tag{6.3}$$

where

$$\rho_n = \frac{n+2}{2(n+1)}\rho_0, \; \varrho_n = \varrho_0^{(\frac{7}{6})^n}, \tag{6.4}$$

and

$$|\boldsymbol{I}_{n+1}^* - \boldsymbol{I}_n^*| \le \varrho_n. \tag{6.5}$$

The initial radii $\rho_0, \varrho_0 < 1$ will be further restricted below. Clearly, ρ_n decreases monotonically toward $\frac{\rho_0}{2}$ as $n \to \infty$. Similarly, $\varrho_n \to 0$ as $n \to \infty$. Moreover, using

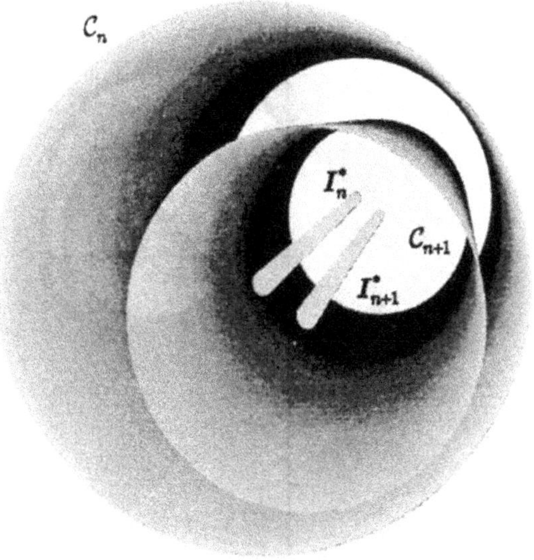

The sequence of high-dimensional complex cylinders.

Eqs. (6.3-6.5) one sees that $C_{n+1} \subset C_n$, $\forall n$, if $\varrho_0 < \left(\frac{3}{4}\right)^6$. It follows that there exists an I_∞^* such that

$$|I_n^* - I_\infty^*| \to 0 \tag{6.6}$$

and the cylinders rapidly shrink to the collapsed complex cylinder

$$C_\infty = \left\{(\varphi, I, q, p) \mid |\Im(\varphi)| \le \frac{\rho_0}{2},\ I = I_\infty^*,\ q = p = 0\right\}. \tag{6.7}$$

Restricted to real angles, this is clearly a torus in state space.

It is further convenient to introduce an intermediate family of cylinders

$$\tilde{C}_n = \left\{(\varphi, I, q, p) \mid |\Im(\varphi)| \le \rho_n - \kappa_n,\ \left|I - I_{n+1}^*\right| \le \varrho_n,\ |q|, |p| \le \varrho_n\right\},\ n \in \mathbb{N} \tag{6.8}$$

where $\kappa_n > 0$. Clearly, $(\varphi, I, q, p) \in \tilde{C}_n$ implies that

$$|I - I_n^*| \le \left|I - I_{n+1}^*\right| + \left|I_{n+1}^* - I_n^*\right| \le 2\varrho_n, \tag{6.9}$$

i.e., $\tilde{C}_n \subset C_n$, for all n. Similarly, $C_{n+1} \subset \tilde{C}_n$, $\forall n$, provided that $\varrho_0 < \frac{1}{6^6}$ and $\kappa_n < \rho_n - \rho_{n+1} \stackrel{def}{=} \Delta\rho_n$.

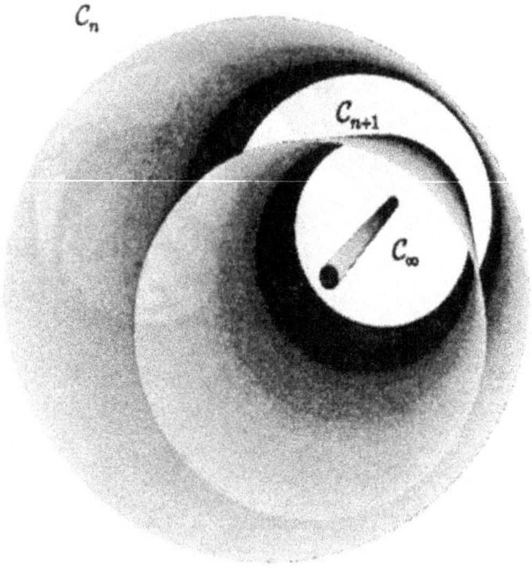

The cylinders converge onto a degenerate cylinder.

Illustration

Let $\kappa_{n+1} = \kappa_n^{\frac{7}{6}}$. Then, since $\kappa_n < \Delta\rho_n$

$$\kappa_{n+1} < \Delta\rho_n^{\frac{7}{6}} = \left(\Delta\rho_n^{\frac{7}{6}}/\Delta\rho_{n+1}\right)\Delta\rho_{n+1} = 3\left(\frac{\rho_0}{4}\right)^{\frac{1}{6}}\Delta\rho_{n+1} \leq \Delta\rho_{n+1}, \tag{6.10}$$

where the last inequality follows provided that $\rho_0 \leq \frac{4}{3^6}$. With the above relation between κ_n and κ_{n+1} we see that

$$\kappa_0 = \varrho_0^\alpha \Rightarrow \kappa_n = \varrho_n^\alpha. \tag{6.11}$$

Thus, fixing ϱ_0 and α immediately fixes κ_n for all n. •

6.2.2 Canonical transformations

The aim of the present discussion is to show the existence of a family of real analytic, canonical coordinate transformations, $\boldsymbol{F}_n$, mapping C_{n+1} onto the subcylinder $\tilde{C}_n$. Since $\tilde{C}_n \subset C_n$, the real analytic composition $\boldsymbol{G}_n = \boldsymbol{F}_0 \circ \boldsymbol{F}_1 \circ \ldots \circ \boldsymbol{F}_n$ maps C_{n+1} into the cylinder $\tilde{C}_0$ for all n. Since the composition of finitely many canonical transformations is canonical, it follows that $\boldsymbol{G}_n$ is also canonical. We would like to show

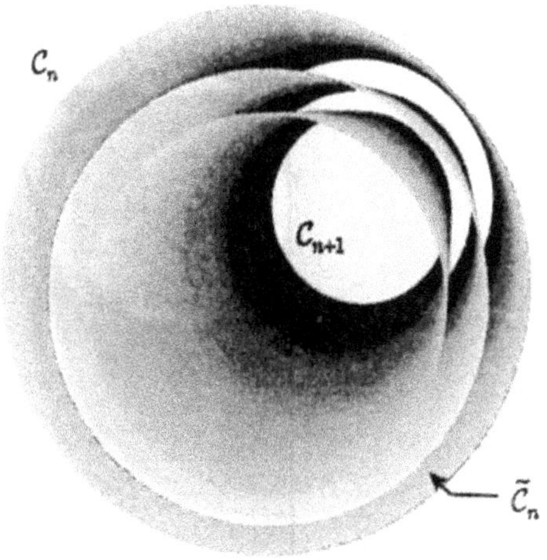

The sequence of interpolating cylinders.

that this sequence of canonical transformations is convergent and that, similarly, the corresponding dynamics converge to a particularly simple flow.

Let $m > n$ and consider the difference

$$\begin{aligned} \|\boldsymbol{G}_n - \boldsymbol{G}_m\|_\infty &= \|\boldsymbol{G}_n - \boldsymbol{G}_n \circ \boldsymbol{F}_{n+1} \circ \ldots \circ \boldsymbol{F}_m\|_\infty \\ &\leq \|\boldsymbol{DG}_n\|_\infty \, \|\boldsymbol{Id} - \boldsymbol{F}_{n+1} \circ \ldots \circ \boldsymbol{F}_m\|_\infty \end{aligned} \tag{6.12}$$

where $\|\cdot\|_\infty = \sup_{\boldsymbol{x} \in \mathcal{C}_\infty} |\cdot|$. Provided $\|\boldsymbol{DG}_n\|_\infty$ is uniformly bounded in n and $\|\boldsymbol{Id} - \boldsymbol{F}_{n+1} \circ \ldots \circ \boldsymbol{F}_m\|_\infty$ can be made arbitrarily small for sufficiently large n and m, then $\{\boldsymbol{G}_n\}$ is a Cauchy sequence in the supremum norm. It follows that $\{\boldsymbol{G}_n\}$ is uniformly convergent to a real analytic map $\boldsymbol{G}_\infty$. In particular, $\boldsymbol{G}_\infty$ provides a, possibly nontrivial, real analytic embedding of $\mathcal{C}_\infty$ in $\mathcal{C}_0$. However, since $\boldsymbol{G}_\infty$ is only defined on the submanifold $\mathcal{C}_\infty$ of $\mathcal{C}_0$, it is not a coordinate transformation and hence not canonical. Nevertheless, we will argue below that a canonical extension of $\boldsymbol{G}_\infty$ does exists on a neighborhood of $\mathcal{C}_\infty$.

The uniform boundedness condition on $\|\boldsymbol{DG}_n\|_\infty$ is easily reformulated to a condition on the individual maps $\boldsymbol{F}_n$. In fact,

$$\|\boldsymbol{DG}_n\|_\infty \leq \prod_{j=0}^{n} \|\boldsymbol{DF}_j\|_\infty , \tag{6.13}$$

where the right-hand side converges if there exists a sequence M_j such that

$$\left|(1 - \|\boldsymbol{DF}_j\|_\infty)\right| \leq M_j, \text{ and } \sum_{j=0}^{\infty} M_j \text{ converges.} \tag{6.14}$$

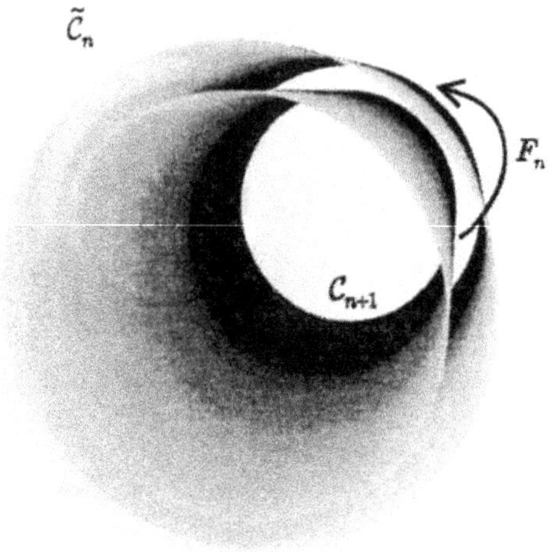

A sequence of canonical transformations viewed in the same space.

Under these conditions, $\|\boldsymbol{DG}_n\|_\infty$ is clearly uniformly bounded and the above results apply. We will indicate below that all these conditions are, in fact, satisfied by our choice of canonical transformations. In particular, it suffices to show that $\boldsymbol{F}_j \to \boldsymbol{Id}$ and $\boldsymbol{DF}_j \to \boldsymbol{I}$ at a sufficiently fast rate as $j \to \infty$.

6.2.3 Dynamics

The above transformations are chosen so that $\{H_n\}$, where

$$H_n(\varphi, \boldsymbol{I}, \boldsymbol{z}) = h_n(\boldsymbol{I}) + \lambda_n(\varphi, \boldsymbol{I})qp + \tilde{H}_n(\varphi, \boldsymbol{I}, \boldsymbol{z}),\ \boldsymbol{z} = (q, p) \tag{6.15}$$

is a family of Hamiltonians that are real analytic on the C_n's, and related through the $\boldsymbol{F}_n$. Moreover, we require that

$$\omega = \boldsymbol{D}_{\boldsymbol{I}} h_n(\boldsymbol{I}_n^*),\ \forall n \tag{6.16}$$

where ω is a constant nonresonant frequency vector satisfying the Diophantine conditions

$$|\boldsymbol{m} \cdot \omega| > \frac{1}{\gamma |\boldsymbol{m}|^\tau},\ \forall \boldsymbol{m} \neq \boldsymbol{0}, \text{ and some } \gamma > 0,\ \tau > n. \tag{6.17}$$

In order to allow sufficiently fast convergence of the canonical transformations $\boldsymbol{G}_n$, it is important to ensure that the supremum $\left\|\tilde{H}_n\right\|_n$ on C_n decreases rapidly enough as $n \to \infty$. In particular, we would like to be able to include terms of $\mathcal{O}\left(\varrho_n^3\right)$ in $\tilde{H}_n$. For example, terms of cubic order in $\boldsymbol{I} - \boldsymbol{I}_n^*$ and $\boldsymbol{z}$ satisfy such an estimate.

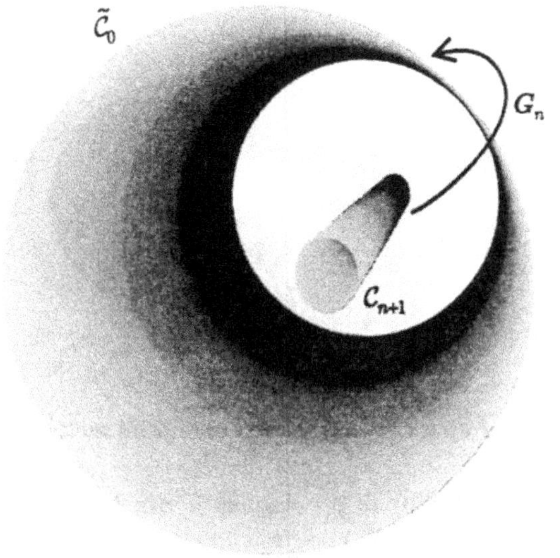

Composing finitely many of the canonical transformations.

This is guaranteed by requiring that

$$\left\|\tilde{H}_n\right\|_n = \mathcal{O}\left(\varrho_n^{\frac{18}{7}}\right), \quad \forall n. \tag{6.18}$$

Furthermore, assuming that h_n, λ_n are uniformly bounded in n, we see that $\{H_n\}$ formally converges to

$$H_\infty(\boldsymbol{\varphi}, \boldsymbol{I}, \boldsymbol{z}) = h_\infty(\boldsymbol{I}) \tag{6.19}$$

on C_∞. However, as mentioned above, the limit map $\boldsymbol{G}_\infty$ is not canonical, since it maps onto a lower dimensional submanifold of the original energy manifold. Hence, the toral flow corresponding to the Hamiltonian H_∞ is not obviously translated to the original flow.

Illustration

Denote the flow associated with the Hamiltonian H_n by $\boldsymbol{\phi}_n(t, \cdot)$. In the limit as $n \to \infty$ the vector field approaches a simple translation in the $\boldsymbol{\varphi}$ direction with speed $\boldsymbol{\omega}$. The resulting motion on C_∞ is then mapped to a nonresonant flow on the embedded torus. We wish to show that this is indeed equivalent to the induced flow of H_0 on the torus.

Consider the difference

$$\boldsymbol{d}(t) = \boldsymbol{\phi}_n(t, (\boldsymbol{\varphi}, \boldsymbol{I}_\infty^*, \boldsymbol{0})) - (\boldsymbol{\varphi} + \boldsymbol{\omega} t, \boldsymbol{I}_\infty^*, \boldsymbol{0}) \tag{6.20}$$

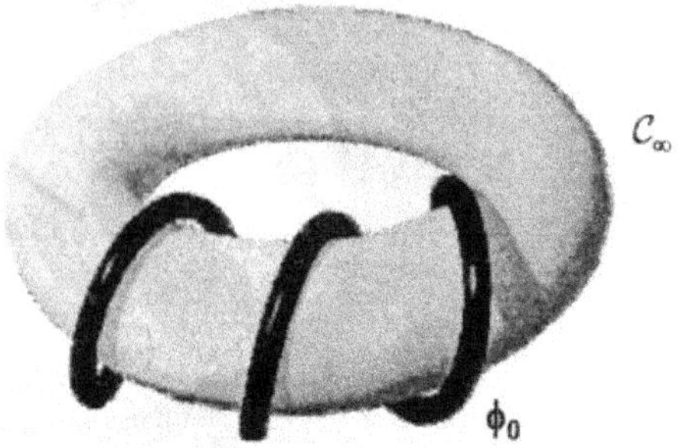

The perturbed vector field converges onto toral motion along C_∞.

for $\varphi \in C_\infty$. Then $\boldsymbol{d}(0) = 0$. Assume it can be shown that $\|\dot{\boldsymbol{d}}\| \leq c\delta(C_{n+1}, \partial C_n)$ on a $\delta(C_{n+1}, \partial C_n)$ neighborhood of C_∞, where c is a constant, independent of n, and $\delta(\cdot,\cdot)$ is the distance between the two sets. Then,

$$|\boldsymbol{d}(t)| \leq \delta(C_{n+1}, \partial C_n) \text{ for } t \in [0, c^{-1}]. \tag{6.21}$$

However, since $\varphi + \omega t \in C_\infty$ for all t, it follows that all points on the line segment between $\varphi + \omega t$ and $\boldsymbol{\phi}_n(t, (\varphi, \boldsymbol{I}^*_\infty, 0))$ lie in C_n for $t \in [0, c^{-1}]$. For $t \in [0, c^{-1}]$ we now find

$$\begin{aligned}
&|\boldsymbol{\phi}_0(t, \boldsymbol{G}_\infty(\varphi, \boldsymbol{I}^*_\infty, 0)) - \boldsymbol{G}_\infty(\varphi + \omega t, \boldsymbol{I}^*_\infty, 0)| \\
&\leq |\boldsymbol{\phi}_0(t, \boldsymbol{G}_\infty(\varphi, \boldsymbol{I}^*_\infty, 0)) - \boldsymbol{\phi}_0(t, \boldsymbol{G}_n(\varphi, \boldsymbol{I}^*_\infty, 0))| \\
&\quad + |\boldsymbol{\phi}_0(t, \boldsymbol{G}_n(\varphi, \boldsymbol{I}^*_\infty, 0)) - \boldsymbol{G}_n(\varphi + \omega t, \boldsymbol{I}^*_\infty, 0)| \\
&\quad + |\boldsymbol{G}_n(\varphi + \omega t, \boldsymbol{I}^*_\infty, 0) - \boldsymbol{G}_\infty(\varphi + \omega t, \boldsymbol{I}^*_\infty, 0)|.
\end{aligned} \tag{6.22}$$

Clearly, the first and last term on the right-hand side go to zero as $n \to \infty$. For the remaining term we find

$$\begin{aligned}
&|\boldsymbol{\phi}_0(t, \boldsymbol{G}_n(\varphi, \boldsymbol{I}^*_\infty, 0)) - \boldsymbol{G}_n(\varphi + \omega t, \boldsymbol{I}^*_\infty, 0)| \\
&= |\boldsymbol{G}_n(\boldsymbol{\phi}_n(t, (\varphi, \boldsymbol{I}^*_\infty, 0))) - \boldsymbol{G}_n(\varphi + \omega t, \boldsymbol{I}^*_\infty, 0)| \\
&\qquad \leq \|\boldsymbol{D}\boldsymbol{G}_n\|_n \, \delta(C_{n+1}, \partial C_n),
\end{aligned} \tag{6.23}$$

where the equality follows from the fact that $\boldsymbol{G}_n$ is a canonical transformation.

Thus, as $n \to \infty$ the right-hand side of Eq. (6.22) goes to zero for $t \in [0, c^{-1}]$ and hence

$$\boldsymbol{\phi}_0(t, \boldsymbol{G}_\infty(\varphi, \boldsymbol{I}^*_\infty, 0)) = \boldsymbol{G}_\infty(\varphi + \omega t, \boldsymbol{I}^*_\infty, 0) \tag{6.24}$$

for $t \in [0, c^{-1}]$. But c does not depend on the initial point, and hence the result holds for all t. We conclude that the perturbed dynamics contain an invariant torus on which the motion is a nonresonant constant angular flow. •

In this and the previous section we have outlined a number of conditions on the map $\boldsymbol{F}_n$ and the family of Hamiltonians $\{H_n\}$. Together these suffice to imply the persistence of invariant tori under sufficiently small perturbations. We now proceed to outline the derivation of the canonical transformations. In particular, we attempt to indicate that they indeed satisfy the necessary conditions for the persistence results to hold.

6.2.4 *The generating function*

To achieve the goal of the previous section we introduce the real analytic generating function

$$S(\varphi, \bar{\boldsymbol{I}}, q, \bar{p}) = \varphi \cdot \bar{\boldsymbol{I}} + q\bar{p} + \tilde{S}(\varphi, \bar{\boldsymbol{I}}, q, \bar{p}). \tag{6.25}$$

Here, barred coordinates lie in $\mathcal{C}_{n+1}$ while the unbarred ones lie in $\tilde{\mathcal{C}}_n$. For $\tilde{H}_n = 0$ the requirements of the previous section are trivially satisfied by $\tilde{S} = 0$. Consequently, for small, but nonzero perturbations, it is reasonable to look for similarly small $\tilde{S}$. A rigorous application of the results of this section requires strict bounds on the magnitudes of the involved quantities. Nevertheless, we shall be satisfied with order of magnitude estimates, since these serve to illustrate the approach without obscuring detail. Suffice it to say that uniform bounds in n are available in place of all $\mathcal{O}(\cdot)$ expressions.

The transformed Hamiltonian now becomes

$$H_{n+1}(\varphi, \bar{\boldsymbol{I}}, q, \bar{p}) = h_n\left(\bar{\boldsymbol{I}} + \tilde{S}_\varphi\right) + \lambda_n\left(\varphi, \bar{\boldsymbol{I}} + \tilde{S}_\varphi\right)\left(\bar{p} + \tilde{S}_q\right) q$$

$$+\tilde{H}_n\left(\varphi, \bar{\boldsymbol{I}} + \tilde{S}_\varphi, q, \bar{p} + \tilde{S}_q\right), \tag{6.26}$$

where subscripts denote partial derivatives and the arguments of $\tilde{S}$ are omitted for simplicity. We wish to determine $\tilde{S}$ so that

$$H_{n+1}(\bar{\varphi}, \bar{\boldsymbol{I}}, \bar{z}) = h_{n+1}(\bar{\boldsymbol{I}}) + \lambda_{n+1}(\bar{\varphi}, \bar{\boldsymbol{I}})\bar{q}\bar{p} + \tilde{H}_{n+1}(\bar{\varphi}, \bar{\boldsymbol{I}}, \bar{z})$$

$$= h_{n+1}(\bar{\boldsymbol{I}}) + \lambda_{n+1}(\varphi + \tilde{S}_{\bar{I}}, \bar{\boldsymbol{I}})(q + \tilde{S}_{\bar{p}})\bar{p} + \tilde{H}_{n+1}(\bar{\varphi}, \bar{\boldsymbol{I}}, \bar{z}), \tag{6.27}$$

where h_{n+1} and λ_{n+1} are close to h_n and λ_n, respectively, on $\mathcal{C}_n$, and $\|\tilde{H}_{n+1}\|_{n+1}$ is much smaller than $\|\tilde{H}_n\|_n$. Equating the two expression for H_{n+1} we obtain

$$\tilde{H}_{n+1}(\bar{\varphi}, \bar{\boldsymbol{I}}, \bar{\boldsymbol{z}}) = h_n\left(\bar{\boldsymbol{I}} + \tilde{S}_\varphi\right) - h_{n+1}(\bar{\boldsymbol{I}})$$

$$+\lambda_n\left(\varphi, \bar{\boldsymbol{I}} + \tilde{S}_\varphi\right)\left(\bar{p} + \tilde{S}_q\right)q - \lambda_{n+1}(\varphi + \tilde{S}_{\bar{I}}, \bar{\boldsymbol{I}})(q + \tilde{S}_p)\bar{p}$$

$$+\tilde{H}_n\left(\varphi, \bar{\boldsymbol{I}} + \tilde{S}_\varphi, q, \bar{p} + \tilde{S}_q\right). \tag{6.28}$$

We proceed to investigate which terms on the right-hand side need to be eliminated in order to achieve the bound on $\tilde{H}_{n+1}$. For ease of notation we introduce the Taylor expansion of $\tilde{S}$ in q and $\bar{p}$:

$$\tilde{S}(\varphi, \bar{\boldsymbol{I}}, q, \bar{p}) = \sum_{j=0}^{\infty} \sum_{k+l=j} \tilde{S}^{kl}(\varphi, \bar{\boldsymbol{I}}) q^k \bar{p}^l. \tag{6.29}$$

Similarly,

$$\tilde{H}_n(\varphi, \bar{\boldsymbol{I}}, q, \bar{p}) = \sum_{j=0}^{\infty} \sum_{k+l=j} \tilde{H}_n^{kl}(\varphi, \bar{\boldsymbol{I}}) q^k \bar{p}^l. \tag{6.30}$$

Further, we recall the Cauchy inequality

$$\left|\frac{\partial^{k_1+k_2+\ldots+k_m} f}{\partial^{k_1} z_1 \partial^{k_2} z_2 \ldots \partial^{k_m} z_m}(\boldsymbol{z}^*)\right| \leq k_1! k_2! \ldots k_m! \frac{M}{r_1^{k_1} r_2^{k_2} \ldots r_m^{k_m}}, \tag{6.31}$$

which we will use to estimate magnitudes of the various terms in Eq. (6.28). Here f is assumed analytic for $|z_j - z_j^*| \leq r_j$ for $j = 1, 2, \ldots, m$ and $M = \max|f(\boldsymbol{z})|$ on $|z_j - z_j^*| = r_j$, $j = 1, 2, \ldots, m$.

We now write the last term in Eq. (6.28) as

$$\tilde{H}_n\left(\varphi, \bar{\boldsymbol{I}} + \tilde{S}_\varphi, q, \bar{p} + \tilde{S}_q\right) = \left[\tilde{H}_n\left(\varphi, \bar{\boldsymbol{I}} + \tilde{S}_\varphi, q, \bar{p} + \tilde{S}_q\right) - \tilde{H}_n\left(\varphi, \bar{\boldsymbol{I}}, q, \bar{p}\right)\right]$$

$$+\tilde{H}_n\left(\varphi, \bar{\boldsymbol{I}}, q, \bar{p}\right) \tag{6.32}$$

and attempt to estimate the magnitude of the bracketed term on C_{n+1}:

$$\left|\tilde{H}_n\left(\varphi, \bar{\boldsymbol{I}} + \tilde{S}_\varphi, q, \bar{p} + \tilde{S}_q\right) - \tilde{H}_n\left(\varphi, \bar{\boldsymbol{I}}, q, \bar{p}\right)\right| \leq \left\|\frac{\partial \tilde{H}_n}{\partial \boldsymbol{I}}\right\| \left\|\tilde{S}_\varphi\right\| + \left\|\frac{\partial \tilde{H}_n}{\partial p}\right\| \left\|\tilde{S}_q\right\|$$

$$= \mathcal{O}\left(\varrho_n^{\frac{11}{7}}\left(\left\|\tilde{S}_\varphi\right\| + \left\|\tilde{S}_q\right\|\right)\right), \tag{6.33}$$

which is $\mathcal{O}\left(\varrho_{n+1}^{\frac{18}{7}}\right)$ if the derivatives of $\tilde{S}$ are $\mathcal{O}\left(\varrho_n^{\frac{10}{7}}\right)$. Assuming this can be achieved, we include the bracketed term in $\tilde{H}_{n+1}$. Furthermore, for the term $\tilde{H}_n^{kl} q^k \bar{p}^l$ in the expansion of $\tilde{H}_n\left(\varphi, \bar{\boldsymbol{I}}, q, \bar{p}\right)$ we find

$$|\tilde{H}_n^{kl} q^k \bar{p}^l| = \mathcal{O}\left(\frac{\varrho_n^{\frac{18}{7}}}{\varrho_n^{k+l}} \varrho_{n+1}^{k+l}\right) = \mathcal{O}\left(\varrho_{n+1}^{\frac{18}{7}\left(\frac{6}{7} + \frac{k+l}{18}\right)}\right) \tag{6.34}$$

on C_{n+1}. This is $\mathcal{O}\left(\varrho_{n+1}^{\frac{18}{7}}\right)$ if $k+l>\frac{18}{7}$. Thus, terms of degree 3 or higher in the expansion of $\tilde{H}_n$ can safely be included in $\tilde{H}_{n+1}$. For simplicity, we will hence assume that $\tilde{S}^{kl}=0$ for $k+l\geq 3$, since these do not affect terms of at most quadratic order in Eq. (6.28).

Illustration

Letting $q=\bar{p}=0$ we express the right-hand side of Eq. (6.28) in the form

$$h_n(\bar{I})-h_{n+1}(\bar{I})+\langle\tilde{H}_n^0\rangle(\bar{I})+\omega\cdot\tilde{S}_\varphi^0-\langle\tilde{H}_n^0\rangle(\bar{I})+\tilde{H}_n^0(\varphi,\bar{I})$$
$$+\left[(D_I h_n(\bar{I})-\omega)\cdot\tilde{S}_\varphi^0\right]$$
$$+\left[h_n(\bar{I}+\tilde{S}_\varphi^0)-h_n(I)-D_I h_n(\bar{I})\cdot\tilde{S}_\varphi^0\right], \tag{6.35}$$

where

$$\langle K\rangle(I)=\frac{1}{(2\pi)^n}\int_{\mathbf{T}^n}K(\varphi,I)\mathrm{d}\varphi \tag{6.36}$$

is the mean of the function K. In order to satisfy the magnitude condition on $\tilde{H}_{n+1}$ it is necessary to eliminate $\left\langle\tilde{H}_n^0\right\rangle$ from the right-hand side. This is simply accomplished by letting

$$h_{n+1}\left(\bar{I}\right)=h_n\left(\bar{I}\right)+\left\langle\tilde{H}_n^0\right\rangle\left(\bar{I}\right). \tag{6.37}$$

Clearly,

$$\|h_{n+1}-h_n\|_n=\mathcal{O}\left(\varrho_n^{\frac{18}{7}}\right).$$

It remains to show that there exists an I_{n+1}^* within ϱ_n from I_n^* so that

$$\omega=D_I h_{n+1}\left(I_{n+1}^*\right). \tag{6.38}$$

This is indeed guaranteed by the implicit function theorem provided that

$$\left|\left(D_I^2 h_n\left(I_n^*\right)\right)^{-1}\right|\left(\frac{\varrho_n^2}{2}\left\|D_I^3 h_n\right\|+\varrho_n\left\|D_I^2\left\langle\tilde{H}_n^0\right\rangle\right\|+\left|D_I\tilde{H}_n^0\left(I_n^*\right)\right|\right)<\frac{\varrho_n}{2} \tag{6.39}$$

and that $D_I^2 h_n$ is invertible for all I within ϱ_n from I_n^* (cf. Chapter 2). The supremum is here taken over a ϱ_n neighborhood of I_n^*. Clearly, condition (6.39) can be satisfied by choosing sufficiently small ϱ_n. In fact, uniform upper bounds in n can be achieved for the inverse:

$$\left|\left(D_I^2 h_n\left(I_n^*\right)\right)^{-1}\right|,$$

thus allowing us to satisfy condition (6.39) at each step of the iterative process by fixing ϱ_0 sufficiently small.

We continue to eliminate terms in Eq. (6.35) by choosing $\tilde{S}^0$ so that

$$\omega \cdot \tilde{S}^0_\varphi - \langle \tilde{H}^0_n \rangle(\bar{I}) + \tilde{H}^0_n(\varphi, \bar{I}) = 0 \tag{6.40}$$

where $\bar{I}$ can be simply thought of as a free parameter. Expanding all terms in Fourier series in φ, we easily solve for the Fourier coefficients of $\tilde{S}^0$ and a unique formal solution is obtained by setting the zeroth order component to zero. Restricted to real angles, convergence of the Fourier series follows, much as in Chapter 2, from the Diophantine conditions. This result can in fact, be extended to $|\Im(\varphi)| < \rho_n - \kappa_n$, provided $\kappa_n > 0$. In particular, one can show that

$$\left\| \tilde{S}^0 \right\| = \mathcal{O}\left(\varrho_n^{\frac{18}{7}} \kappa_n^{1-2d} \right) \tag{6.41}$$

on this complex domain.

With this choice for $\tilde{S}^0$ it is now possible to show that, for sufficiently small κ_0 and ϱ_0, the remaining terms in Eq. (6.35) can all be made small enough to be included in $\tilde{H}_{n+1}$ and need not concern us further. •

We can now write the right-hand side of Eq. (6.28) as

$$\begin{aligned}
\omega \cdot \sum_{j=1}^{2} \sum_{k+l=j} \tilde{S}^{kl}_\varphi + \lambda_n(\varphi, \bar{I})(\bar{p} + \tilde{S}_q)q - \lambda_{n+1}(\varphi, \bar{I})(q + \tilde{S}_{\bar{p}})\bar{p} + \sum_{j=1}^{2} \sum_{k+l=j} \tilde{H}^{kl}_n(\varphi, \bar{I}) q^k \bar{p}^l \\
+ \left[(D_I h_n(\bar{I}) - \omega) \cdot \tilde{S}_\varphi \right] + \left[h_n(\bar{I} + \tilde{S}_\varphi) - h_n(\bar{I}) - D_I h_n(\bar{I}) \cdot \tilde{S}_\varphi \right] \\
\left[\lambda_{n+1}(\varphi, \bar{I})(q + \tilde{S}_{\bar{p}})\bar{p} - \lambda_{n+1})\varphi + \tilde{S}_{\bar{I}}, \bar{I})(q + \tilde{S}_{\bar{p}})\bar{p} \right] \\
\left[\lambda_n(\varphi, \bar{I} + \tilde{S}_\varphi)(\bar{p} + \tilde{S}_q)q - \lambda_n(\varphi, \bar{I})(\bar{p} + \tilde{S}_q)q \right].
\end{aligned} \tag{6.42}$$

One immediately sees that the coefficient of $q\bar{p}$ in the first line may be made to vanish by setting $\tilde{S}^{11} = 0$ and

$$\lambda_{n+1}(\varphi, I) = \lambda_n(\varphi, I) + \tilde{H}^{11}_n(\varphi, I), \tag{6.43}$$

from which it immediately follows that

$$\|\lambda_{n+1} - \lambda_n\|_n = \mathcal{O}\left(\varrho_n^{\frac{4}{7}} \right). \tag{6.44}$$

As in the illustration, we proceed by choosing $\tilde{S}^{kl}$ so as to eliminate all terms in the first line. It should come as no surprise that the remaining bracketed terms can then be made sufficiently small to be included in $\tilde{H}_{n+1}$ by appropriate choices of ϱ_0 and κ_0.

Illustration

The remaining coefficients $\tilde{S}^{kl}$ can be shown to satisfy equations of the form

$$\omega \cdot u_\varphi(\varphi) + \lambda(\varphi)u(\varphi) = \tilde{H}(\varphi) \tag{6.45}$$

for $|\Im(\varphi)| \leq \rho_n$ and given functions λ and $\tilde{H}$. Along the curve $\varphi(t) = \varphi(0) + \omega t$ we obtain the ordinary differential equation

$$\frac{du}{dt}(\varphi(t)) + \lambda(\varphi(t))u(\varphi(t)) = \tilde{H}(\varphi(t)) \tag{6.46}$$

with the general solution

$$u(\varphi(t)) = e^{-\int_0^t \lambda(\varphi(t'))dt'}\left[u(\varphi(0)) + \int_0^t \tilde{H}(\varphi(s))e^{\int_0^s \lambda(\varphi(t'))dt'}ds\right]. \tag{6.47}$$

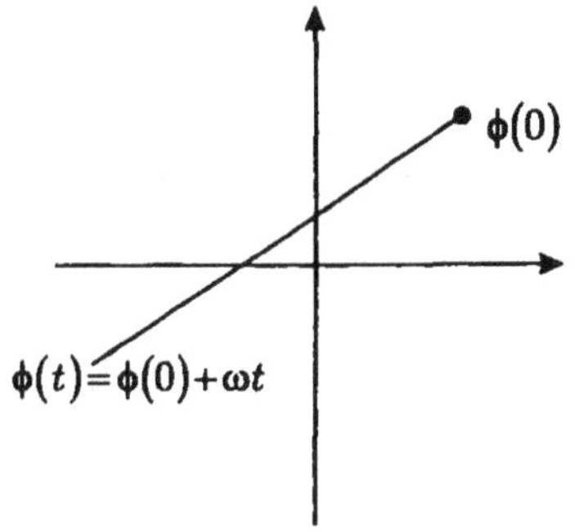

The partial differential equation is integrated along characteristics.

We now assume that $\lambda \geq \tilde{\lambda} > 0$. Then, the exponential factor blows up as $t \to -\infty$. Thus, bounded solutions are possible only if

$$u(\varphi(0)) = \int_{-\infty}^0 \tilde{H}(\varphi(s))e^{\int_0^s \lambda(\varphi(t'))dt'}ds, \tag{6.48}$$

which gives the solution at an arbitrary point $\varphi(0)$. Moreover,

$$\|u\| \leq \|\tilde{H}\| \int_{-\infty}^0 e^{-\int_s^0 \tilde{\lambda}dt'}ds = \frac{\|\tilde{H}\|}{\tilde{\lambda}}, \tag{6.49}$$

which can be used to estimate magnitudes of the $\tilde{S}^{kl}$. Note that if $\lambda \leq \tilde{\lambda} < 0$ the discussion applies instead to $t \to \infty$. •

6.2.5 *Implications*

We have discussed how a generating function may be constructed at each step in the iteration procedure. A more careful examination of the various steps involved allows one to conclude that, in fact, the canonical transformation $\boldsymbol{F}_n$ and its Jacobian $\boldsymbol{DF}_n$, corresponding to the generating function S, rapidly approach the identity map and the unit matrix as required for the existence of $\boldsymbol{G}_\infty$.

In addition to requiring that the frequency vector satisfy the Diophantine conditions, it was necessary to assume the invertibility of $\boldsymbol{D}_I^2 h_n$ on $\tilde{C}_n$. Furthermore, λ_n was required to be bounded away from zero. Assuming that these conditions are satisfied for $n = 0$, it is possible to show that uniform results in n can be achieved by using Eqs. (6.37) and (6.43) iteratively.

It follows that the family of canonical transformation $\{\boldsymbol{F}_n\}$ can be shown to satisfy all the conditions stipulated in Sections 6.2.1-6.2.2. In particular, one has thereby shown the persistence of invariant tori with nonresonant flow under sufficiently small perturbations of the Hamiltonian $h_0(\boldsymbol{I}) + \lambda_0 qp$.

Illustration

One drawback of the above discussion is that the limit map $\boldsymbol{G}_\infty$ was not a canonical transformation, since it maps to a lower dimensional submanifold of the original manifold. We have been able to show the presence of invariant tori in the perturbed flow with nonresonant constant flow, but no information has been given on the flow in the vicinity of the tori. It turns out that such information can, in fact, be extracted from the sequence of canonical transformations $\boldsymbol{G}_n$.

Consider truncating the generating function corresponding to $\boldsymbol{G}_n$ at second order. The resulting transformed Hamiltonian is identical to that obtained from the full transformation apart from terms of $\mathcal{O}(3)$. One can show that the corresponding sequence of canonical transformations also converges to $\boldsymbol{G}_\infty$ on C_∞. However, since the generating functions are quadratic, it is not necessary to let their domains of definition shrink down to C_∞. Instead, they can be shown to converge to a generating function yielding a canonical transformation on a neighborhood of C_∞. Consequently, the transformed Hamiltonian becomes

$$H_\infty(\varphi, \boldsymbol{I}, \boldsymbol{z}) = \boldsymbol{\omega} \cdot (\boldsymbol{I} - \boldsymbol{I}_\infty^*) + \frac{1}{2}(\boldsymbol{I} - \boldsymbol{I}_\infty^*) \cdot \boldsymbol{D}_I^2 h_\infty(\boldsymbol{I}_\infty^*)(\boldsymbol{I} - \boldsymbol{I}_\infty^*)$$

$$+\lambda_\infty(\varphi, \boldsymbol{I}) qp + \mathcal{O}(|\boldsymbol{I} - \boldsymbol{I}_\infty^*|^3, |\boldsymbol{z}|^3) \tag{6.50}$$

on some neighborhood of C_∞. In particular, λ_∞ is bounded away from zero. Obviously, the invariant torus corresponds to $\boldsymbol{I} = \boldsymbol{I}_\infty^*$, $\boldsymbol{z} = \boldsymbol{0}$. Furthermore, neglecting higher order terms we find that the $q = 0$ and $p = 0$ submanifolds are also invariant, and correspond to exponential *contraction and expansion respectively. Thus, the local dynamics of the unperturbed tori appear to be preserved in the perturbed flow.* •

6.3 General conclusions

In the previous section we studied a particular unperturbed Hamiltonian and the effect of small perturbations on certain invariant tori. We now suggest that this result actually applies to a more general class of Hamiltonians with similar basic state-space structure. Consider the Hamiltonian

$$H(\varphi, \boldsymbol{I}, \boldsymbol{z}; \varepsilon) = H_0(\boldsymbol{I}, \boldsymbol{z}) + \varepsilon H_1(\varphi, \boldsymbol{I}, \boldsymbol{z}; \varepsilon) \tag{6.51}$$

(subscripts should not be confused with those used in the previous section), where H_0 and H_1 are real analytic in their arguments on the cylinder $\mathcal{C}_0$ and $\boldsymbol{z} = (q, p)$. It is assumed that

$$\boldsymbol{D}_I H_0(\boldsymbol{I}_0^*, \boldsymbol{0}) = \boldsymbol{\omega}, \tag{6.52}$$

where $\boldsymbol{\omega}$ satisfies the Diophantine conditions. Furthermore, we assume that

$$(\boldsymbol{D}_z H_0)(\boldsymbol{I}, \boldsymbol{0}) = \boldsymbol{0} \tag{6.53}$$

for all $\boldsymbol{I}$ in some region $\mathcal{V}$, and that the eigenvalues of the linearization $(\boldsymbol{J}\boldsymbol{D}_z^2 H_0)(\boldsymbol{I}, \boldsymbol{0})$ at the origin lie away from the imaginary axis.

On $\mathcal{C}_0$ terms of third or higher order in q, p and $\boldsymbol{I} - \boldsymbol{I}_0^*$ will all be $\mathcal{O}\left(\varrho_0^{\frac{18}{7}}\right)$. These can therefore be combined with εH_1 to $\tilde{H}_0$ provided ε is sufficiently small. Thus, the results of the previous section will apply, provided we can find a transformation of the (q, p) coordinates such that the Taylor expansion including terms of degree two will have the form $h(\boldsymbol{I}) + \lambda q p$, where $\lambda \neq 0$. Taylor expanding H_0 we obtain

$$H_0(\boldsymbol{I}, q, p) = \boldsymbol{\omega} \cdot (\boldsymbol{I} - \boldsymbol{I}_0^*) + \frac{1}{2}(\boldsymbol{I} - \boldsymbol{I}_0^*) \cdot \boldsymbol{D}_I^2 H_0(\boldsymbol{I}_0^*, \boldsymbol{0})(\boldsymbol{I} - \boldsymbol{I}_0^*) + \frac{1}{2}\boldsymbol{z} \cdot \boldsymbol{D}_z^2 H_0(\boldsymbol{I}_0^*, \boldsymbol{0})\boldsymbol{z} \tag{6.54}$$

plus terms of third order or higher. Consider the linear coordinate transformation

$$\boldsymbol{z} = \boldsymbol{Q}\bar{\boldsymbol{z}}, \tag{6.55}$$

which is canonical provided $\det \boldsymbol{Q} = 1$. The desired form of the Hamiltonian then is obtained if $\boldsymbol{Q}$ can be found so that

$$\boldsymbol{Q}^T \boldsymbol{D}_z^2 H_0(\boldsymbol{I}_0^*, \boldsymbol{0}) \boldsymbol{Q} = \begin{pmatrix} 0 & \lambda \\ \lambda & 0 \end{pmatrix} \tag{6.56}$$

or, equivalently,

$$\boldsymbol{Q}^{-1} \boldsymbol{J} \boldsymbol{D}_z^2 H_0(\boldsymbol{I}_0^*, \boldsymbol{0}) \boldsymbol{Q} = \begin{pmatrix} \lambda & 0 \\ 0 & -\lambda \end{pmatrix} \tag{6.57}$$

i.e., Q diagonalizes $JD_z^2 H_0(I_0^*, 0)$. By the assumption on the eigenvalues of $JD_z^2 H_0(I_0^*, 0)$, such a Q clearly exists. Thus, our assertion is confirmed and we conclude that nonresonant tori of the unperturbed system corresponding to H_0 survive the perturbation for sufficiently small ε.

Illustration

It is, in fact, possible to prove a much stronger claim regarding the geometry of the flow near the persistent tori. In the unperturbed system, the invariant tori are associated with stable and unstable manifolds on which orbits approach and leave the tori, respectively, exponentially fast. In the transformed system these are simply invariant manifolds tangent at the torus to the $q = 0$ and $p = 0$ planes.

As in the case of hyperbolic stationary points, it is reasonable to ask whether these manifolds survive the perturbation along with their corresponding tori. In fact, it is possible to show the existence of a real analytic canonical transformation $(\varphi, I, z) \longmapsto (\bar{\varphi}, \bar{I}, \bar{z})$, such that on the manifold $\bar{I} = \bar{I}_0$, $\bar{p} = 0$ the flow is given by

$$\frac{d}{dt}\bar{I} = \frac{d}{dt}\bar{p} = 0 \tag{6.58}$$

and

$$\frac{d}{dt}\bar{\varphi} = \omega, \quad \frac{d}{dt}\bar{q} = f(\bar{\varphi}, \bar{q}; \varepsilon)\, \bar{q}, \tag{6.59}$$

where $f(\bar{\varphi}, \bar{q}; \varepsilon) = f_0 + \mathcal{O}(\varepsilon)$, and f_0 is a nonzero constant. In other words, the $\bar{I} = \bar{I}_0$, $\bar{p} = 0$ is invariant and the flow on it approaches $\bar{q} = 0$ exponentially as $t \to \infty$ or $-\infty$, depending on the sign of f_0. We further note that on the persistent manifolds the frequency in the transformed angle variables is still ω. •

If the stable and unstable manifolds of the stationary point in the z degree-of-freedom coincide to form a homoclinic loop, then the corresponding stable and unstable manifolds of an invariant torus also coincide in a homoclinic manifold. Of course, similar conclusions hold in the case of a heteroclinic connection in the z degree-of-freedom. Our experience from previous chapters suggests that emphasis should be put on the fate of these manifolds under small perturbations. In the illustration we have argued that stable and unstable manifolds of persistent tori survive small perturbations. It is not very likely that they continue to coincide. Instead, we expect the manifolds to separate, leaving behind points of intersection just as in the planar case.

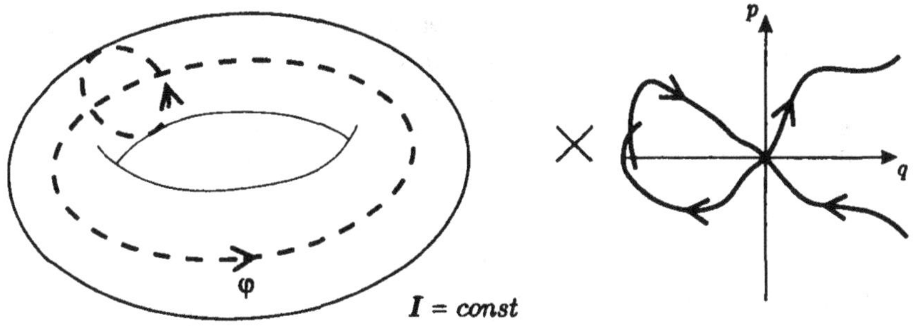

While the perturbed system no longer decouples, the surviving tori have stable and unstable manifolds which may intersect.

We will explore some of the consequences of such persistent intersections toward the end of this chapter and further in the application in Chapter 7. But first we embark on a journey similar to that in Chapter 4, with the goal of obtaining computable conditions for the existence of transversal intersections of the surviving manifolds.

6.4 The Perturbation Approach

In this section we attempt to find approximate expressions for solutions to the perturbed Hamiltonian system that lie in the stable and unstable manifolds of persistent tori. As discussed above, we assume that the unperturbed manifolds coincide to form a homo- or heteroclinic manifold. The approach is that of Chapter 4 and, as we shall see, there are many analogies with the treatment there. Due to the higher dimensionality of the present discussion, however, additional conditions have to be imposed on the perturbation solutions.

6.4.1 The ansatz

We denote solutions on the perturbed manifolds by

$$\boldsymbol{y}_\epsilon(t,t_0) = (\boldsymbol{\varphi}_\epsilon(t,t_0), \boldsymbol{I}_\epsilon(t,t_0), \boldsymbol{z}_\epsilon(t,t_0))^T .$$

We assume that these solutions can be written as expansions in powers of ϵ as follows:

$$\boldsymbol{y}_\epsilon(t,t_0) = \boldsymbol{y}_0(t-t_0) + \sum_{k=1}^{k=N} \varepsilon^k \boldsymbol{y}_k(t,t_0) + o\left(\varepsilon^N\right). \tag{6.60}$$

Here $\boldsymbol{I}_0(t,t_0) \equiv \boldsymbol{I}_0(t_0,t_0) =$ constant, $\boldsymbol{z}_0(t-t_0)$ is a solution to the unperturbed system for fixed value of $\boldsymbol{I} = \boldsymbol{I}_0$ on the homo- or heteroclinic manifold, and

$$\boldsymbol{\varphi}_0(t-t_0) = \int_{t_0}^{t} \frac{\partial H_0}{\partial \boldsymbol{I}}(\boldsymbol{I}_0, \boldsymbol{z}_0(s-t_0))\, ds + \boldsymbol{\varphi}_0(0).$$

The choice of integer N depends on the degree of accuracy that one wishes to achieve. We note that any expression involving functions of φ_0, or the arbitrary parameters that are introduced at each step in the perturbation analysis, depends on the vector of parameters $\varphi_0(0)$. Allowing variations in t_0, this vector serves to map out the entire unperturbed manifold and therefore allows us to consider perturbed solutions based at different points of the unperturbed manifold. The $\varphi_0(0)$ dependence will in most cases be suppressed in order to avoid obscuring the results.

We require that the above expansions are valid on semi-infinite time intervals. In particular, all involved functions should be bounded as $t \to \pm\infty$ for orbits on the stable and unstable manifolds, respectively. Since the dynamics are asymptotic to motion on the persistent invariant tori, they should all approach quasiperiodic functions. This will, in turn, impose restrictions on any free parameters that arise in our method. We refer to the coefficients of ε^k in the above expansions as the k-th corrections to the unperturbed solutions. Unless otherwise stated, single subscripts on vector quantities refer to a vector of corrections of a particular order, whereas in the case of double subscripts, the first index denotes individual components in this vector.

6.4.2 *The variational equation*

In order to obtain differential equations for the k-th order corrections, we substitute the expansion (6.60) into the equations of motion corresponding to Eq. (6.51) and identify coefficients of equal powers of ε. The resulting system of equations for the k-th correction is called the k-th variational equation and has the following general form:

$$\begin{pmatrix} \dot{\varphi}_k \\ \dot{I}_k \\ \dot{z}_k \end{pmatrix} = \begin{pmatrix} D_z D_I H_0(z_0(t-t_0), I_0) & 0 & D_I^2 H_0(z_0(t-t_0), I_0) \\ 0 & 0 & 0 \\ J D_z^2 H_0(z_0(t-t_0), I_0) & 0 & D_I J D_z H_0(z_0(t-t_0), I_0) \end{pmatrix} \begin{pmatrix} \varphi_k \\ I_k \\ z_k \end{pmatrix}$$
$$+ \begin{pmatrix} g_k^{\varphi}\left(y_0(t-t_0), y_1(t,t_0), \ldots, y_{k-1}(t,t_0)\right) \\ g_k^{I}\left(y_0(t-t_0), y_1(t,t_0), \ldots, y_{k-1}(t,t_0)\right) \\ g_k^{z}\left(y_0(t-t_0), y_1(t,t_0), \ldots, y_{k-1}(t,t_0)\right) \end{pmatrix}. \tag{6.61}$$

Rather than include the lower-order corrections in the argument of the last term of this equation, we simply write $g_k^{\cdot}(t, t_0)$ below.

A general solution

We note that each variational equation is a system of linear, nonhomogeneous, ordinary differential equations with time-dependent coefficients. In particular, the coefficients are known, provided that an unperturbed solution has been chosen, and the nonhomogeneities are determined once the lower-order corrections have been solved for. Moreover, due to the assumed form of the unperturbed Hamiltonian system, there are no derivatives with respect to the φ variables present in the above equations. Consequently, the I_k equation decouples from the remaining equations and can be solved independently at each stage in the perturbation process. We write the solution as

$$I_k(t,t_0) = \delta_k(t_0) + \int_{t_0}^{t} g_k^I(s,t_0)\,\mathrm{d}s. \tag{6.62}$$

Here $\delta_k(t_0)$ is a vector of constants corresponding to the initial values of I_k. At this stage it provides n free parameters to be specified when choosing a particular orbit.

Substituting Eq. (6.62) into the last component of Eq. (6.61), we obtain

$$\dot{q}_k = \frac{\partial^2 H_0}{\partial q \partial p}(I_0, z_0(t-t_0))\,q_k + \frac{\partial^2 H_0}{\partial p^2}(I_0, z_0(t-t_0))\,p_k + g_k^q(t,t_0) \tag{6.63}$$

and

$$\dot{p}_k = -\frac{\partial^2 H_0}{\partial q^2}(I_0, z_0(t-t_0))\,q_k - \frac{\partial^2 H_0}{\partial q \partial p}(I_0, z_0(t-t_0))\,p_k + g_k^p(t,t_0). \tag{6.64}$$

This is a system of two linear, nonhomogeneous, ordinary differential equations in two unknowns. We will assume that these equations are nondegenerate, in as far as there exist two independent homogeneous solutions. Just as in Chapter 4, a complete solution can be obtained if one homogeneous solutions can be found. A second homogeneous solution can then be derived by the method of reduction of order, and a particular solution follows from the method of variation of parameters.

We adapt the expressions of Chapter 4 to the present situation. One homogeneous solution is trivially obtained from the unperturbed solution:

$$(\phi_{11}(t-t_0), \phi_{12}(t-t_0))^T = \left(\frac{\partial H^0}{\partial p}(I_0, z_0(t-t_0)), -\frac{\partial H^0}{\partial q}(I_0, z_0(t-t_0))\right)^T. \tag{6.65}$$

If $\phi_{11}(t_0,t_0) \neq 0$ then a second homogeneous solution is given by

$$\phi_{21}(t-t_0) = \phi_{11}(t-t_0)\int_{t_0}^{t} \frac{\partial^2 H_0}{\partial p^2}(I_0, z_0(t-t_0))\,[\phi_{11}(s-t_0)]^{-2}\,\mathrm{d}s, \tag{6.66}$$

and

$$\phi_{22}(t-t_0) = \phi_{12}(t-t_0)\int_{t_0}^{t} \frac{\partial^2 H_0}{\partial p^2}(I_0, z_0(t-t_0))\,[\phi_{11}(s-t_0)]^{-2}\,\mathrm{d}s + [\phi_{11}(t-t_0)]^{-1}, \tag{6.67}$$

over some interval containing $t = t_0$. Similarly, if, instead, $\phi_{12}(t_0,t_0) \neq 0$, then the functions

$$\phi_{21}(t-t_0) = \phi_{11}(t-t_0)\int_{t_0}^{t} \frac{\partial^2 H_0}{\partial q^2}(I_0, z_0(t-t_0))\,[\phi_{12}(s-t_0)]^{-2}\,\mathrm{d}s - [\phi_{12}(t-t_0)]^{-1} \tag{6.68}$$

and

$$\phi_{22}(t-t_0) = \phi_{12}(t-t_0) \int_{t_0}^{t} \frac{\partial^2 H_0}{\partial q^2} (\boldsymbol{I}_0, \boldsymbol{z}_0(t-t_0)) \left[\phi_{12}(s-t_0)\right]^{-2} \mathrm{d}s \qquad (6.69)$$

provide a second homogeneous solution. The particular choice of second homogeneous solution presented above assures that $\phi_{11}\phi_{22} - \phi_{21}\phi_{12} \equiv 1$ for all time.

Using the fundamental matrix

$$\boldsymbol{Z}(t,t_0) = \begin{pmatrix} \phi_{11}(t-t_0) & \phi_{21}(t-t_0) \\ \phi_{12}(t-t_0) & \phi_{22}(t-t_0) \end{pmatrix}, \qquad (6.70)$$

the complete solution to Eqs. (6.63-6.64) is

$$\boldsymbol{z}_k(t,t_0) = \boldsymbol{Z}(t,t_0) \left[\begin{pmatrix} \alpha_k(t_0) \\ \beta_k(t_0) \end{pmatrix} + \int_{t_0}^{t} \boldsymbol{Z}^{-1}(s,t_0) \boldsymbol{g}_k^{z}(s,t_0) \mathrm{d}s \right]. \qquad (6.71)$$

The constants $\alpha_k(t_0)$ and $\beta_k(t_0)$ are at this point free parameters that need to be determined when choosing a particular orbit. If we substitute the known solutions for the corrections $\boldsymbol{z}_k$ and $\boldsymbol{I}_k$ into the φ component of Eq. (6.61), we obtain a set of uncoupled differential equations with completely known right-hand sides (apart from the free parameters $\alpha_k(t_0)$, $\beta_k(t_0)$ and $\boldsymbol{\delta}_k(t_0)$). The solutions can then be written

$$\varphi_k(t,t_0) = \boldsymbol{\gamma}_k(t_0) + \int_{t_0}^{t} \left[\begin{array}{c} \boldsymbol{D}_z \boldsymbol{D}_I H_0(\boldsymbol{I}_0, \boldsymbol{z}_0(s-t_0))\, \boldsymbol{z}_k(s,t_0) + \\ \boldsymbol{D}_I^2 H_0(\boldsymbol{I}_0, \boldsymbol{z}_0(s-t_0))\, \boldsymbol{I}_k(s,t_0) + \boldsymbol{g}_k^{\varphi}(s,t_0) \end{array} \right] \mathrm{d}s, \qquad (6.72)$$

where $\boldsymbol{\gamma}_k(t_0)$ is yet another vector of free parameters, which we will shortly specify. Eqs. (6.62) and (6.71-6.72) now provide a complete solution to the k-th variational equation, provided that the lower-order corrections are known. One can continue in this way to obtain the perturbation expansion to arbitrary order.

6.4.3 The stable and unstable manifolds

It remains, however, to determine the $2n+2$ free parameters given by $\alpha_k(t_0)$, $\beta_k(t_0)$, $\boldsymbol{\gamma}_k(t_0)$, and $\boldsymbol{\delta}_k(t_0)$. In Chapter 4 this choice was partially effected by restricting attention to a direction normal to the unperturbed manifold. Similarly, we consider the hyperplane

$$\Pi = \operatorname{span}\left\{ \left(0,0,\ldots,0,0, \frac{\partial H_0}{\partial q}(\boldsymbol{I}_0, \boldsymbol{z}_0(0)), \frac{\partial H_0}{\partial p}(\boldsymbol{I}_0, \boldsymbol{z}_0(0))\right)^T, \boldsymbol{e}_{I_1}, \ldots, \boldsymbol{e}_{I_n} \right\}$$

based at $\boldsymbol{y}_0(0)$, where $\boldsymbol{e}_{I_i}$ denotes a unit vector in the I_i direction. The tangent space, T, of the unperturbed homoclinic manifold at $\boldsymbol{y}_0(0)$ is spanned by the vector $\boldsymbol{J}\boldsymbol{D}_z H_0(\boldsymbol{I}_0, \boldsymbol{z}_0(0))$ and n unit vectors in the φ directions. Clearly, $T \oplus \Pi = \mathbb{R}^{2n+2}$, since T is orthogonal to Π and their dimensions add up to the dimensionality of phase space. It follows that the unperturbed manifolds intersect Π transversely at the point $\boldsymbol{y}_0(0)$. The transversality of the intersection implies its persistence under small perturbations. In particular, the perturbed stable and unstable manifolds each intersect Π at at least one point.

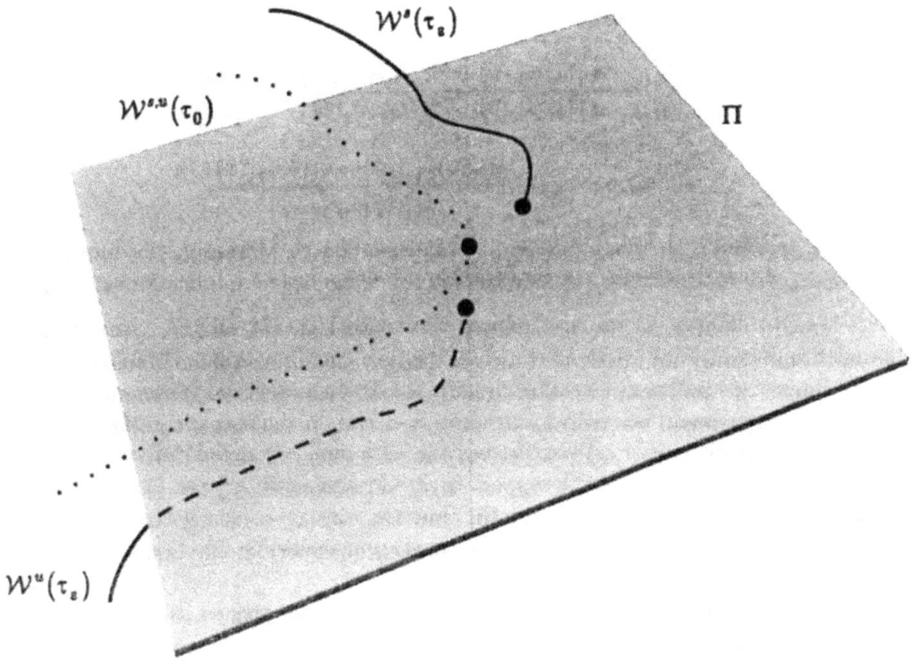

The perturbed stable and unstable manifolds intersect the plane Π at isolated points.

Illustration

It is possible that the perturbed stable manifold, say, intersects Π at more than one point. However, since an unperturbed solution intersects Π only once, the perturbation expansion can only capture one of the intersections. Essentially, this amounts to restricting attention to the point whose forward or backward orbit, respectively, never intersects Π again. •

We shall therefore require that the k-th correction

$$\begin{aligned} y_k(t_0, t_0) &= (\gamma_k(t_0), \delta_k(t_0), \alpha_k(t_0)\phi_{11}(0) + \beta_k(t_0)\phi_{21}(0), \\ &\quad \alpha_k(t_0)\phi_{12}(0) + \beta_k(t_0)\phi_{22}(0))^T \end{aligned} \quad (6.73)$$

be parallel to this plane and thus that the perturbation expansion intersects this plane at $t = t_0$, independently of order of truncation. This criterion will be referred to as the intersection condition.

Using Eq. (6.65) we find that

$$\frac{q_k(t_0,t_0)}{p_k(t_0,t_0)} = \frac{\alpha_k(t_0)\,\phi_{11}(0)+\beta_k(t_0)\,\phi_{21}(0)}{\alpha_k(t_0)\,\phi_{12}(0)+\beta_k(t_0)\,\phi_{22}(0)} = -\frac{\phi_{12}(0)}{\phi_{11}(0)} \Rightarrow$$

$$\alpha_k(t_0) = -\beta_k(t_0)\,\frac{\phi_{12}(0)\phi_{22}(0)+\phi_{11}(0)\phi_{21}(0)}{\phi_{11}^2(0)+\phi_{12}^2(0)}. \tag{6.74}$$

Thus, $\alpha_k(t_0)$ and $\beta_k(t_0)$ cannot be selected independently. Moreover, we immediately see that $\boldsymbol{\gamma}_k(t_0)\equiv \mathbf{0}$, whereas the intersection condition has no implications on $\boldsymbol{\delta}_k(t_0)$.

As in Chapter 4, we now impose the condition that all the corrections be bounded in forward and backward time on the stable and unstable manifolds, respectively. Since the stable and unstable manifolds follow an identical treatment through a simple time reversal, we restrict attention to orbits on the stable manifold.

The inhomogeneity $\boldsymbol{g}_k^I$ can be written as a sum of partial derivatives of H, with at least one derivative with respect to $\boldsymbol{\varphi}$. Consequently, $\boldsymbol{g}_k^I$ can be written as a function that approaches a quasiperiodic function with no constant term as $t\to\infty$. It follows that both sides of Eq. (6.62) approach quasiperiodic functions, since there are no secular terms in the integrand.

The properties of the ϕ_{ij} follow in much the same way as in Chapter 4. In particular, one finds that ϕ_{11} and ϕ_{12} both remain bounded as $t\to\infty$. However, ϕ_{21} is seen to blow up in forward time. Thus, in the notation of Chapter 4, $\beta_k(t_0)$ must cancel the contribution from $-\int_{t_0}^{t} c_{2k}(s,t_0)\,ds$ for large t. This is accomplished by letting

$$\beta_k(t_0) = \int_{t_0}^{\infty} c_{2k}(s,t_0)ds, \tag{6.75}$$

and, consequently, uniquely specifying $\alpha_k(t_0)$ through Eq. (6.74). With these choices one finds that q_k and p_k both approach quasiperiodic functions as $t\to\infty$.

It remains to fix $\boldsymbol{\delta}_k(t_0)$. To this end we will require that no constant terms be present on the right-hand side of the differential equation for $\boldsymbol{\varphi}_k$ as $t\to\infty$. This will be referred to as the no-secular-terms condition. From the $\boldsymbol{\varphi}$ component of Eq. (6.61), we see that such constant terms stem from the constant part of the quasiperiodic functions that the various terms approach in forward time. By setting the constant term in the equations for $\dot{\varphi}_{ik}$, $i=1,\ldots,n$, equal to zero, a linear system in the n unknown components of $\boldsymbol{\delta}_k(t_0)$ is obtained. This can be uniquely solved provided that $\det\left[\boldsymbol{D}_{\boldsymbol{I}}^2 H_0(\tau_0(\boldsymbol{I}_0))\right]\neq 0$, which of course is already required of the persistent tori.

With the unique choices made above, all the free parameters of the solutions on the perturbed manifolds are fixed and particular orbits thus chosen. Since the manifolds intersect the plane Π transversely, it follows that an appropriate measure of the separation of the manifolds is obtained by considering the distance between these points in the plane Π. This task will be considered in the following section.

Illustration

No use has yet been made of the Hamiltonian character of the perturbed flow. As we shall show shortly, it implies that knowledge of $\boldsymbol{\delta}_i(t_0)$ for $i = 1, \ldots, k$ and $\beta_j(t_0)$ for $j = 1, \ldots, k-1$ uniquely specifies the value of $\beta_k(t_0)$. In this way one can inductively show that the $2n+2$ arbitrary parameters that appear at each order in the general solution are uniquely determined, once $\boldsymbol{\delta}_k(t_0)$ is fixed for all orders up to and including the present.

Since the perturbed system is still Hamiltonian, orbits on the perturbed manifolds have the same value of the Hamiltonian, H, as does the corresponding perturbed torus, $\tau_\varepsilon(\boldsymbol{I}_0)$. Thus,

$$H(\boldsymbol{y}_\varepsilon(t_0, t_0); \varepsilon) - H(\tau_\varepsilon(\boldsymbol{I}_0)) = 0. \tag{6.76}$$

If we expand in terms of powers of ϵ and set the coefficients of each separate power equal to zero, one finds

$$\frac{\partial H_0}{\partial q}(\boldsymbol{I}_0, \boldsymbol{z}_0(0))\left(\alpha_k(t_0)\phi_{11}(0) + \beta_k(t_0)\phi_{21}(0)\right)$$

$$+\frac{\partial H_0}{\partial p}(\boldsymbol{I}_0, \boldsymbol{z}_0(0))\left(\alpha_k(t_0)\phi_{12}(0) + \beta_k(t_0)\phi_{22}(0)\right) + \boldsymbol{D}_I H_0(\boldsymbol{I}_0, \boldsymbol{z}_0(0)) \cdot \boldsymbol{\delta}_k(t_0)$$

$$+\tilde{H}_k + F_k(\alpha_1(t_0), \beta_1(t_0), \delta_1(t_0), \ldots, \alpha_{k-1}(t_0), \beta_{k-1}(t_0), \delta_{k-1}(t_0)) = 0, \tag{6.77}$$

where F is some function of the given arguments, $\alpha_i(t_0)$ has to satisfy Eq. (6.74), and

$$\tilde{H}_k = \frac{1}{k!}\frac{d^k}{d\varepsilon^k} H(\tau_\epsilon(\boldsymbol{I}_0))$$

is a constant for a given torus. Using Eq. (6.65), one finds that the coefficient in front of $\beta_k(t_0)$ in the above equation is $\phi_{11}(0)\phi_{22}(0) - \phi_{21}(0)\phi_{12}(0)$, which $= 1$. Thus, we can solve inductively for $\beta_k(t_0)$ in terms of $\boldsymbol{\delta}_i(t_0)$ and $\tilde{H}_i$, $i = 1, \ldots, k$, where the latter is uniquely specified by choosing a particular torus or, equivalently, the corresponding value of $\boldsymbol{I}_0$ on the unperturbed torus. •

The conclusion of the illustration is particularly natural in light of the discussion in Chapter 4. In particular, the two equivalent measures of manifold separation derived there correspond to distances in the $\boldsymbol{z}$ and $\boldsymbol{I}$ degrees-of-freedom, respectively. As we shall see shortly, these are equal to differences in $\beta_k(t_0)$ and $\boldsymbol{\delta}_k(t_0)$, respectively, and their equivalence follows from the illustration above.

6.5 The Manifold Separation

We are now in a position to consider the splitting of the manifolds, as described by the distance between the points of intersection of the hyperplane Π and the stable and unstable manifolds of the perturbed tori, respectively. For clarity we will add the superscripts s and u to any quantity on the stable and unstable manifolds, respectively.

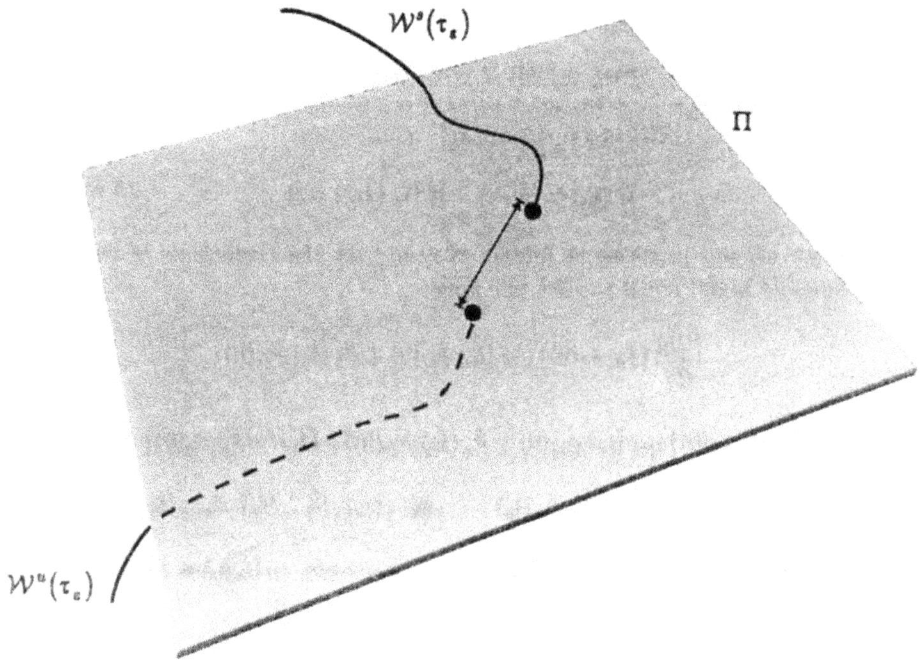

We measure the separation of the manifolds along Π.

From the expression for $y_k(t_0, t_0)$ we immediately see that $\varphi_k^s(t_0, t_0) = \varphi_k^u(t_0, t_0)$. Similarly, the projection of the separation vector onto the line spanned by

$$(0, \ldots, 0, \frac{\partial H_0}{\partial q}(I_0, z_0(0)), \frac{\partial H_0}{\partial p}(I_0, z_0(0)))$$

is given by $\beta_k^s(t_0) - \beta_k^u(t_0) = \Delta\beta_k(t_0)$. Finally, along each of the remaining n vectors spanning Π, the separation between the points is $\Delta\delta_k(t_0) = \delta_k^s(t_0) - \delta_k^u(t_0)$. However, we showed above that $\beta_k^{s,u}(t_0)$ was uniquely determined once $\delta_i^{s,u}(t_0)$, $i = 1, \ldots, k$, and the value of I_0 on the unperturbed torus were fixed. Since I_0 is the same for two unperturbed tori connected by a homo- or heteroclinic manifold, it follows that if $\Delta\delta_i(t_0)$ vanishes for $i = 1, \ldots, k$, then $\Delta\beta_k(t_0)$ is automatically zero.

Illustration

In the two-degree-of-freedom case, we found that the methods of Chapter 4 applied after a reduction to a periodically perturbed single-degree-of-freedom Hamiltonian system. In particular, the perturbed Hamiltonian was equivalent to the perturbed action of the original system. It was found that the manifold separation was equally well described by the distance in the z direction as by the difference in the Hamiltonian on the perturbed manifolds. In light of the discussion in the present chapter, the conclusions of Chapter 4 are simply special cases of the more general relation between $\Delta\beta_k(t_0)$ and $\Delta\boldsymbol{\delta}_k(t_0)$. •

If the perturbed manifolds were to intersect at a point, then by the standard theorems on uniqueness of solutions they would have to intersect along an entire orbit. It is straightforward to show that the points on such a curve would be parametrized by the parameter t_0. Consequently, it suffices to consider the manifold separation for $t_0 = 0$. In particular, using the implicit function theorem we can now conclude that if $\Delta\boldsymbol{\delta}_i(0) \equiv \mathbf{0}$ for $i = 1, \ldots, k-1$ and if, for a particular choice of $\varphi_0(0)$, it is true that

1. $\Delta\boldsymbol{\delta}_k(0)$ vanishes, and
2. $\boldsymbol{D}_{\varphi_0(0)}\Delta\boldsymbol{\delta}_k(0)$ is nonsingular,

then the perturbed manifolds intersect transversely.

In the homoclinic case, the evaluation of the manifold separation is particularly simple. As can be seen from the first variational equation for φ_k, the constant contribution to the right-hand side as t approaches $\pm\infty$ is the same for the stable and unstable manifolds, as long as they both approach the same perturbed torus, since the asymptotic motion is dense on the torus. Thus, in the homoclinic case, the determination of $\Delta\boldsymbol{\delta}_1(0)$ boils down to finding the constant part of

$$\int_0^t \frac{\partial H_1}{\partial \varphi}(\varphi_0(s), \boldsymbol{I}_0, z_0(s))\, ds$$

as $t \to \pm\infty$. Alternatively, one can arrive at the same conclusion from Eq. (6.62) by observing that as the perturbed torus is approached, the left-hand sides approach quasiperiodic functions with the same constant part. The result then follows from applying this fact to the right-hand sides.

In the heteroclinic case, the constant terms are no longer the same in forward and backward time. Consequently, one has to include the contribution of the other terms in the variational equation for φ_k such as $\boldsymbol{D}_z\boldsymbol{D}_I H_0(\boldsymbol{I}, z_0(t))\, z_k(t)$ and $\boldsymbol{g}_k^{\varphi}(t)$. In the general situation, the above integral is, therefore, no longer sufficient for the determination of the distance between the manifolds.

6.6 Consequences of transversal intersections - Arnold diffusion

6.6.1 A special case of Arnold's example

We consider the perturbed three-degree-of-freedom Hamiltonian

$$H_\varepsilon(\varphi, \boldsymbol{I}, \boldsymbol{z}) = \frac{1}{2}\left(p^2 + I_1^2\right) + I_2 - (1 - \cos q) + \varepsilon (1 - \cos q) f(q, \varphi), \tag{6.78}$$

such that $(q, \varphi) \in \mathbb{T}^3$ and $(p, \boldsymbol{I}) \in \mathbb{R}^3$. Here f is periodic with period 2π in its arguments. We restrict attention to the $H_\varepsilon = 0$ submanifold, since this Hamiltonian is simply the suspension of the Hamiltonian obtained by omitting I_2 and setting $\varphi_2 = t$. In particular, $\dot\varphi_2 = 1$, i.e., $\varphi_2(t) = t + \varphi_2(0)$, for all ε. Moreover, the submanifold $\boldsymbol{z} = 0$ is invariant under the full flow and the perturbation vanishes on this submanifold.

In the unperturbed problem, $\varepsilon = 0$, the various degrees-of-freedom decouple completely, thus rendering a completely integrable Hamiltonian. In particular, the action variables $\boldsymbol{I}$ are constants of the motion, and the $\boldsymbol{z}$ degree-of-freedom contains two homoclinic orbits based at $q = 0 (= 2\pi)$. The homoclinic motion is given by

$$p_0(t) = \frac{2}{\cosh t}, \quad q_0(t) = 4 \arctan e^t. \tag{6.79}$$

Thus, the unperturbed dynamics foliate the $\boldsymbol{z} = \boldsymbol{0}$ submanifold with invariant two-dimensional tori, each of which is connected to itself through two three-dimensional homoclinic manifolds. Since $H_\varepsilon \equiv 0$, it follows that the invariant tori are parameterized by the value of I_1 alone.

As mentioned above, due to the form of the perturbation, all the invariant tori persist for arbitrary ε, as do their stable and unstable manifolds. Using the theory of the previous sections we have argued that transversal intersections of the perturbed manifolds correspond to zeros of the functions $\Delta\boldsymbol{\delta}_1(0)$ such that the Jacobian $\boldsymbol{D}_{\varphi_0(0)}\Delta\boldsymbol{\delta}_1(0)$ is nonsingular. In particular, $\Delta\boldsymbol{\delta}_1(0)$ was shown to be simply evaluated by finding the difference between the constant parts of the quasiperiodic functions $\int_0^t \frac{\partial H_1}{\partial \varphi}(\varphi_0(s), \boldsymbol{I}_0, \boldsymbol{z}_0(s))\, ds$ as $t \to \pm\infty$. In the case of Eq. (6.78), however, the integrand is easily seen to decay exponentially fast to zero, and thus the integral is absolutely convergent. Consequently,

$$\Delta\boldsymbol{\delta}_1(0) = \int_{-\infty}^{\infty} \frac{\partial H_1}{\partial \varphi}(\varphi_0(s), \boldsymbol{I}_0, \boldsymbol{z}_0(s))\, ds. \tag{6.80}$$

Finally, the condition on the Jacobian implies that the manifolds intersect transversely if the functional

$$F_I(\varphi_0(0)) = \int_{-\infty}^{\infty} (1 - \cos q_0(s)) f(q_0(s), \varphi_0(s))\, ds \tag{6.81}$$

has a nondegenerate extremum for some $\varphi_0(0)$.

Assuming the existence of such an extremum for $I_{10}(0) \in [\frac{1}{2}, \frac{5}{2}]$, it is possible to prove the following claim using variational analysis. In particular, for small ε, consider a perturbed orbit such that

$$q(0) = -\pi,\ q(T) = \pi,\ \varphi(0) = \varphi^0,\ \varphi(T) = \varphi^T \tag{6.82}$$

on a given interval $[0, T]$. Provided that $\varphi_2^T - \varphi_2^0 = T \geq \frac{A}{\varepsilon}$ for some constant $A = \mathcal{O}(1)$, then there is a unique such orbit which lies close to the unperturbed separatrix

$$q_0(t),\ \boldsymbol{I} = \boldsymbol{I}_0,\ \varphi_1(t) = I_1 t + \varphi_{10}(0) \tag{6.83}$$

on the interval $\left[0, \frac{|\ln \varepsilon|}{6}\right]$, to the unperturbed torus $\tau_0(\boldsymbol{I}_0)$ on $\left[\frac{|\ln \varepsilon|}{6}, T - \frac{|\ln \varepsilon|}{6}\right]$, and finally to the unperturbed separatrix

$$q_0(t - T),\ \boldsymbol{I} = \boldsymbol{I}_0,\ \varphi_1(t) = I_1 t + \varphi_{10}(0) \tag{6.84}$$

on $\left[T - \frac{|\ln \varepsilon|}{6}, T\right]$.

The shadowing orbits lie initially along the stable manifold, wind around the torus and then escape along an unstable manifold.

We have thus constructed a shadowing orbit for which $\boldsymbol{I} \approx \boldsymbol{I}_0^0$, connecting the boundary points given in Eq. (6.82). Since all the invariant tori survive the perturbation, it follows that there are stable and unstable manifolds of tori with $\boldsymbol{I}_0^1 \neq \boldsymbol{I}_0^0$ arbitrary close to those belonging to $\tau(\boldsymbol{I}_0^0)$. A natural question is then whether the shadowing orbit obtained in the previous discussion can be concatenated to a neighboring shadowing orbit, thus setting up dynamics that drift over large distances in the action variables. In particular, we would like the resulting transition orbits to persist for $\varepsilon \to 0$.

In fact, the variational method can be extended further to show the existence of an orbit connecting the neighborhoods of two separate tori, $\tau\left(I_1'\right)$ and $\tau\left(I_1''\right)$, such that $I_1' - I_1'' \leq \varepsilon$ for sufficiently small ε. In particular, the connecting orbit is the concatenation of two shadowing orbits constructed on the intervals $[0, T']$ and $[T', T'']$ such that $T'' = 2\frac{A}{\varepsilon}$, respectively. Continuing this process of concatenating the shadowing orbits, we obtain orbits connecting neighborhoods of arbitrarily separated tori (provided $I_1 \in \left[\frac{1}{2}, \frac{5}{2}\right]$), thus allowing large changes in the action variables even in the limit of vanishing ε. Moreover, the typical time scale for this drift is bounded above by $\frac{A'}{\varepsilon^2}$ for some $A' = \mathcal{O}(1)$.

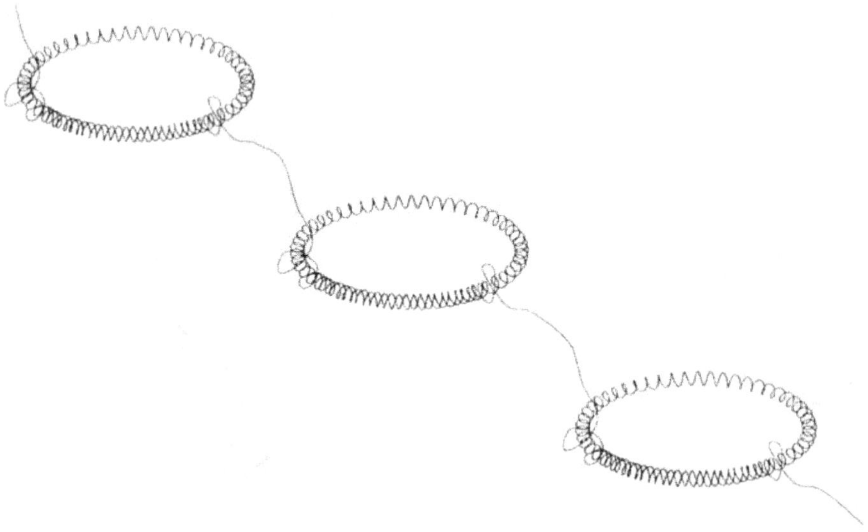

Large excursions are possible by concatenating many shadowing orbits.

The induced instability of the persistent tori is known as **Arnold diffusion**, cf. Chapter 2. Contrary to the situation in Chapter 2, it is only polynomially slow in the perturbation parameter and could thus potentially have physical implications compared to other time scales of a physical problem. We will return to this in the next chapter. But first we need to consider the more general situation discussed in the first part of this chapter, in which only a fraction of the invariant tori survive the perturbation.

6.6.2 Transition chains in arbitrary applications

The example in the previous section was particularly suitable for studying Arnold diffusion, since the perturbation vanished on the invariant tori. Consequently, the tori all persisted under the perturbation and none of the issues encountered in previous sections had to be considered. One notable effect of this was that perturbed tori could be found arbitrarily close independently of ε, allowing one to construct excursive orbits as concatenations of orbits shadowing the tori's stable and unstable manifolds. In the more general case, for fixed ε, perturbed invariant tori would typically be separated by a finite distance. It would no longer necessarily be possible to have stable and unstable manifolds of separate tori approach each other sufficiently close for concatenations to exist.

Consider a curve segment in action space with full torsion; i.e., such that the tangent vector and its $n-1$ first derivatives span $\mathbb{R}^n$ for all points on the curve. The results in the first sections of this chapter can be generalized to show that the maximum gap between surviving tori is less than $c\varepsilon^a$ for some positive constant c and an estimable number $a \in (0,1)$. The perturbed invariant manifolds of the surviving tori are assumed to intersect transversely. From the discussion on the manifold separation, we see that the angle between the intersecting manifolds is typically $\mathcal{O}(\varepsilon)$. It follows that for the stable and unstable manifolds of separate tori to intersect, these tori should be within $\mathcal{O}(\varepsilon)$ from each other. Clearly, this is generally not possible, since $a < 1$.

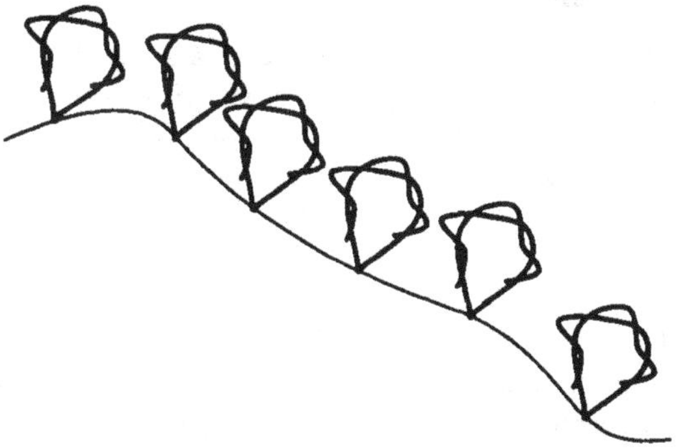

In general, the separation between surviving tori might be too large to allow the concatenation discussed above.

The solution to this quandary lies in the possibility of pushing the perturbation up to $\mathcal{O}(\varepsilon^{p+1})$ for some integer p. This can actually be achieved provided that the frequency vector has no resonances up to order p, or in other words that

$$\boldsymbol{D_I} H_0(\boldsymbol{I}) \cdot \boldsymbol{m} \neq 0, \text{ for all } \boldsymbol{m} \in \mathbb{Z}^n, \ \sum_{i=1}^{n} |m_i| \leq p \tag{6.85}$$

in some neighborhood of the curve segment. Through a series of canonical transformations, the perturbed Hamiltonian can then be transformed to the form

$$\bar{H}_0(\boldsymbol{I}, \boldsymbol{z}; \varepsilon) + \varepsilon^{p+1}\bar{H}_1(\boldsymbol{\varphi}, \boldsymbol{I}, \boldsymbol{z}; \varepsilon) \tag{6.86}$$

where the $\bar{H}_0$ term satisfies the condition on the unperturbed Hamiltonian stipulated in Section 6.1.3. It follows that ε in the above estimate of the gap between surviving tori can be replaced by ε^p. Consequently, if $ap > 1$, then, for ε sufficiently small, the gap size allows neighboring stable and unstable manifolds to intersect, thus forming the transition chain on which the drift orbits are built.

Provided the nonresonance condition can be satisfied, one is likely to have connections of shadowing orbits that permit large excursions in the action variables. As we shall see in the next chapter, this makes it possible for orbits to exist which sample dramatically different dynamical behaviors. In the presence of lower order resonances, it is generically not possible to establish such intersections. Whether this prevents the existence of connecting orbits is not clear. After all, the gaps in action space are partially due to resonant unperturbed tori that were destroyed by the perturbation. One might well imagine processes by which the remnants of these tori would act to set up additional connections between persistent tori. This will not, however, be further discussed in this book.

6.7 Notes and references

A mixture of results dating back to the late sixties and developments from this past year, this chapter has considered the generalization of the ideas introduced in earlier chapters to many degrees-of-freedom. The discussion of the persistent tori and their stable and unstable manifolds is a condensation of the work of Moser[1] and Graff[2]. In particular, the choice of detail and illustrations has been made to enable the reader to digest the latter paper with minor additional effort. The analysis of the perturbed manifolds in terms of power series expansions has appeared in several sources, most notably the paper by Chierchia & Gallavotti[3], which also discusses the KAM results for the persistent tori in much detail. The particular formulation chosen in this chapter is further discussed in Dankowicz[4]. The compendium of Lochak[5] and the paper by Bernard[6] contain the results on Arnold diffusion discussed above. Lochak also discusses the development of novel methods for dealing with Arnold diffusion in the original KAM case of Chapter 2.

Very much related to the present discussion is the early work of Holmes & Marsden[7], later generalized by Wiggins[8] and made rigorous by Robinson[9]. Naturally, the higher-dimensional Melnikov vector discussed by these authors is equivalent to the manifold separation considered here. However, these authors consider evaluating the separation by choosing a sequence of integration limits for the conditionally convergent integrals considered above, rather than focusing on their asymptotic means.

1. J. Moser, "Convergent Series Expansions for Quasiperiodic Motions, *Mathematische Annalen* **169** (1967), pp. 136-176.
2. S. M. Graff, "On the Conservation of Hyperbolic Invariant Tori for Hamiltonian Systems," *Journal of Differential Equations* **15** (1974), pp. 1-69.
3. L. Chierchia and G. Gallavotti, "Drift and Diffusion in Phase Space," *Annales de l'IHP, Section Physique Théorique* **60** (1994), pp. 1-144.
4. H. Dankowicz, "Analytical Expressions for Stable and Unstable Manifolds in Higher Degree of Freedom Hamiltonian Systems," *International Journal of Bifurcation and Chaos* **6(11)** (1996), pp. 1997-2013.
5. P. Lochak, "Arnold Diffusion: A Compendium of Remarks and Questions," in *Hamiltonian Systems with Three of More Degrees of Freedom*, NATO Adv. Sci. Inst. C Math. Phys. Sci. (Kluwer, Dordrecht, 1996).
6. P. Bernard, "Perturbation d'un Hamiltonien Partiellement Hyperbolique," *Comptes Rendus de l'Académie des Sciences Paris, Série I. Mathématique,* **t. 323,** (1996), pp. 189-194.
7. P. J. Holmes and J. E. Marsden, "Melnikov's Method and Arnold Diffusion for Perturbations of Integrable Hamiltonian Systems," *Journal of Mathematical Physics* **23**(4) (1982), pp. 669-675.
8. S. Wiggins, *Global Bifurcations and Chaos* (Springer Verlag, 1988).
9. C. Robinson, "Horseshoes for Autonomous Hamiltonian Systems using the Melnikov Integral," *Ergodic Theory & Dynamical Systems* **8*** (1988), pp. 395-409.

Chapter 7
APPLICATION - RADIATION PRESSURE PROBLEMS IN CELESTIAL MECHANICS. PART II

Science is climactic. As dreary as the necessary analysis of a given phenomena into its many components may be, their eventual successful synthesis should not leave even the most passing observer unscathed. That nature allows useful knowledge to be derived from this process of dissection and reassembly, rather than its opposite, is a mystery. That the total, nevertheless, is greater than the sum of its parts is an equally accepted proposition with similar existential implications.

Up to this point, this book has, piece by piece, gathered a vast collection of tools and techniques, as well as qualitative information about the complex behavior of particular classes of problems. We are in the possession of all the ingredients necessary to approach an actual physical problem and draw conclusions that are, in some sense, measurable. This synthesis is but a modest example of the explanatory power contained in the discussions of the previous chapters. We will return to some of its other potential applications toward the end of this book.

In this chapter we again focus on the radiation pressure problem introduced in Chapter 5. In particular, we study the dynamical features present when the third degree-of-freedom is included. The results of the previous chapter are shown to apply, and, specifically, the concept of Arnold diffusion is suggested as a means for particles in certain regions of state space to escape the asteroid. Of particular importance is estimating the rate at which this occurs, since it bears relevance on the expected amount of grains in orbit about asteroids.

In Section 1 we present a generalization of the Levi-Civita transformation of Chapter 5, which requires the general treatment of canonical transformations given in Chapter 2. This transformation is applied to a three-body problem with radiation pressure in Section 2, with a Hamiltonian of the form studied in the previous chapter. In Sections 3 and 4 we describe the properties of the unperturbed and perturbed dynamics along the lines used in Chapter 6. Finally, Sections 5 and 6 contain actual computations of the manifold separation, and an estimate of the rate of Arnold diffusion and its physical implications.

7.1 The canonical Kustaanheimo-Stiefel transformation

In Chapter 5 we used the Levi-Civita transformation to simplify our Hamiltonian and remove the singularity at the origin. In this section we will describe its higher-dimensional analogue. In particular, we consider a coordinate transformation $\mathbb{R}^8 \to \mathbb{R}^6$, which satisfies the Poisson bracket conditions in Chapter 2. Note that the additional degree-of-freedom makes it likely that the bracket conditions are not satisfied for all points in state space, but possibly only on restricted submanifolds. Indeed, this is the case.

Consider the matrix

$$\Lambda(\boldsymbol{u}) = \begin{pmatrix} u_1 & -u_2 & -u_3 & u_4 \\ u_2 & u_1 & -u_4 & -u_3 \\ u_3 & u_4 & u_1 & u_2 \end{pmatrix} \tag{7.1}$$

where $\boldsymbol{u} \in \mathbb{R}^4$. The Kustaanheimo-Stiefel (KS) transformation is then given by

$$\boldsymbol{q} = \Lambda(\boldsymbol{u})\boldsymbol{u} \tag{7.2}$$

and

$$\boldsymbol{p} = \frac{1}{2r}\Lambda(\boldsymbol{u})\boldsymbol{w} \tag{7.3}$$

where $\boldsymbol{q}$, $\boldsymbol{p}$ and $\boldsymbol{u}$, $\boldsymbol{w}$ are pairs of conjugate variables. Here $r = |\boldsymbol{u}|^2 = |\boldsymbol{q}|$. It is convenient to introduce the notation

$$\boldsymbol{v}^* = (v_4, -v_3, v_2, -v_1)\,, \tag{7.4}$$

where $\boldsymbol{v}$ is any vector in $\mathbb{R}^4$. Clearly, $\boldsymbol{v}$ and $\boldsymbol{v}^*$ are orthogonal under the standard scalar product in $\mathbb{R}^4$. We define the bilinear operator $l : \mathbb{R}^4 \times \mathbb{R}^4 \to \mathbb{R}$ by $l(\boldsymbol{w}, \boldsymbol{u}) = \boldsymbol{w} \cdot \boldsymbol{u}^*$.

We proceed to check the Poisson bracket conditions. It is straightforward to show that

$$\{q_i, q_j\} = 0, \text{ and } \{q_i, p_j\} = \delta_{ij}. \tag{7.5}$$

Further,

$$\{p_i, p_j\} = \frac{1}{2r^3} l(\boldsymbol{w}, \boldsymbol{u}) \begin{pmatrix} 0 & q_3 & -q_2 \\ -q_3 & 0 & q_1 \\ q_2 & -q_1 & 0 \end{pmatrix} \tag{7.6}$$

and the KS transformation is clearly only canonical on the submanifold $l(\boldsymbol{w}, \boldsymbol{u}) = 0$. Thus, restricting attention to this submanifold, we simply substitute $\boldsymbol{q}(\boldsymbol{u})$ and $\boldsymbol{p}(\boldsymbol{u}, \boldsymbol{w})$ from above into a given Hamiltonian H to obtain

$$\bar{H}(\boldsymbol{u}, \boldsymbol{w}, u_0, w_0) = H(\boldsymbol{q}(\boldsymbol{u}), \boldsymbol{p}(\boldsymbol{u}, \boldsymbol{w}), q_0, p_0), \text{ where } u_0 = q_0, \text{ and } w_0 = p_0, \tag{7.7}$$

and we have

$$\frac{d\boldsymbol{u}}{ds} = \frac{\partial \bar{H}}{\partial \boldsymbol{w}}, \text{ and } \frac{d\boldsymbol{w}}{ds} = -\frac{\partial \bar{H}}{\partial \boldsymbol{u}}.$$

It is necessary to show that $l(\boldsymbol{w}, \boldsymbol{u})$ is invariant under the flow of $\bar{H}$, in order for the canonical formalism to hold for all time. But this follows from the fact that

$$\frac{dl(\boldsymbol{w},\boldsymbol{u})}{ds} = \frac{\partial l(\boldsymbol{w},\boldsymbol{u})}{\partial \boldsymbol{w}} \cdot \frac{d\boldsymbol{w}}{ds} + \frac{\partial l(\boldsymbol{w},\boldsymbol{u})}{\partial \boldsymbol{u}} \cdot \frac{d\boldsymbol{u}}{ds} = l(\frac{d\boldsymbol{w}}{ds}, \boldsymbol{u}) + l(\boldsymbol{w}, \frac{d\boldsymbol{u}}{ds})$$

$$= l(-\frac{\partial \bar{H}}{\partial \boldsymbol{u}}, \boldsymbol{u}) + l(\boldsymbol{w}, \frac{\partial \bar{H}}{\partial \boldsymbol{w}}) = \frac{\partial H}{\partial q_j} l(\boldsymbol{u}, \frac{\partial q_j}{\partial \boldsymbol{u}}) + \frac{\partial H}{\partial p_j}\left[l(\boldsymbol{u}, \frac{\partial p_j}{\partial \boldsymbol{u}}) + l(\boldsymbol{w}, \frac{\partial p_j}{\partial \boldsymbol{w}})\right] \tag{7.8}$$

in fact vanishes identically. Thus, the KS transformation restricted to the $l(\boldsymbol{w}, \boldsymbol{u}) = 0$ submanifold is a canonical transformation from $\mathbb{R}^8$ to $\mathbb{R}^6$.

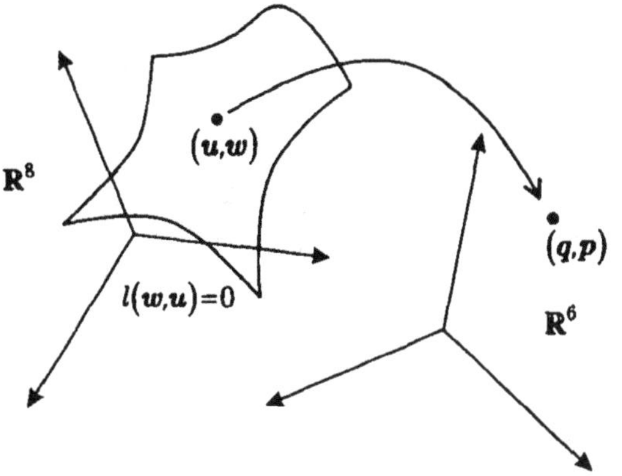

The KS transformation is canonical on the invariant submanifold $l\,(\boldsymbol{w}, \boldsymbol{u}) = 0$.

Illustration

We consider applying the KS transformation to the integrable two-body problem discussed in Chapter 1:

$$H(\boldsymbol{q}, \boldsymbol{p}) = \frac{1}{2}|\boldsymbol{p}|^2 - \frac{\mu}{r}. \tag{7.9}$$

As before, we rescale the independent variable by

$$\frac{dt}{ds} = r, \tag{7.10}$$

which yields

$$H_s(\boldsymbol{q}, \boldsymbol{p}, p_0) = \frac{r}{2}|\boldsymbol{p}|^2 - \mu + r p_0, \tag{7.11}$$

where $H_s \equiv 0$. The KS transformation yields

$$r|\boldsymbol{p}|^2 = \frac{1}{4}|\boldsymbol{w}|^2 - \frac{1}{4r}l^2(\boldsymbol{w}, \boldsymbol{u}) = \frac{1}{4}|\boldsymbol{w}|^2, \tag{7.12}$$

where the last equality follows from restricting attention to the $l(\boldsymbol{w}, \boldsymbol{u}) = 0$ submanifold. Thus, we obtain

$$\bar{H}_s(\boldsymbol{w}, \boldsymbol{u}, w_0) = \frac{1}{8}|\boldsymbol{w}|^2 + w_0 r, \tag{7.13}$$

where $\bar{H}_s \equiv \mu$. Since u_0 is cyclic, we find $w_0 \equiv$ constant $= -E$, where E is the energy of the motion.

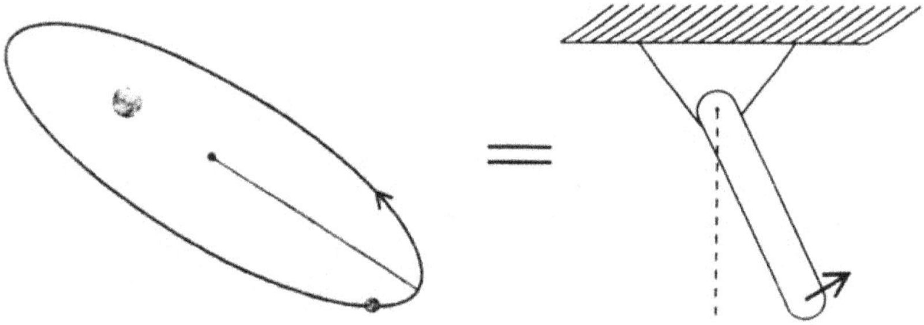

Bounded two-body dynamics are equivalent to those of a linear pendulum.

We now obtain the canonical equations of motion

$$\ddot{\boldsymbol{u}} - \frac{E}{2}\boldsymbol{u} = 0, \tag{7.14}$$

i.e., the equations of a harmonic oscillator with frequency $\sqrt{-E/2}$. As already shown in Chapter 1, bounded motion corresponds to $E < 0$. The pleasant fact that the KS transformation turns the equations into those of a harmonic oscillator suggests its use in perturbed two-body problems where the resulting equations would have the appearance of nonlinear oscillators. •

We finally discuss the process of choosing initial conditions in the $\boldsymbol{u}$, $\boldsymbol{w}$ coordinates given similar conditions in the original coordinates. In particular, it is necessary to initially satisfy $l(\boldsymbol{w}, \boldsymbol{u}) = 0$ in order for the above results to apply. Clearly, there are many vectors $\boldsymbol{u}(0)$ satisfying

$$\boldsymbol{q}(0) = \Lambda(\boldsymbol{u}(0))\boldsymbol{u}(0). \tag{7.15}$$

Assume that one such vector has been chosen. Then solving Eq. (7.3) together with $l(\boldsymbol{w}, \boldsymbol{u}) = 0$ yields

$$\boldsymbol{w}(0) = 2\Lambda^T(\boldsymbol{u}(0))\boldsymbol{p}(0) \tag{7.16}$$

and the initial conditions for the momentum variables are uniquely determined once the positions are specified.

Illustration

We briefly discuss the inverse image in $\mathbb{R}^4$ of a point in $\mathbb{R}^3$. In fact, from the first component of the KS transformation, it follows that

$$u_1^2 + u_4^2 = \frac{1}{2}(r + q_1), \text{ and } u_2^2 + u_3^2 = \frac{1}{2}(r - q_1), \tag{7.17}$$

i.e., circles in the u_1, u_4 and u_2, u_3 planes. We thus write

$$(u_1, u_4) = \sqrt{\frac{1}{2}(r + q_1)}(\cos\alpha, \sin\alpha) \tag{7.18}$$

and

$$(u_2, u_3) = \sqrt{\frac{1}{2}(r - q_1)}(\cos\beta, \sin\beta) \tag{7.19}$$

where α and β are arbitrary angles. Substitution into the remaining components of the KS transformation yields

$$\cos\beta = \frac{q_2\cos\alpha + q_3\sin\alpha}{2\sqrt{q_2^2 + q_3^2}}, \text{ and } \sin\beta = \frac{q_3\cos\alpha - q_2\sin\alpha}{2\sqrt{q_2^2 + q_3^2}}, \tag{7.20}$$

uniquely fixing β as a function of α.

We thus have a curve in $\mathbb{R}^4$ parameterized by α, such that the entire curve is mapped onto the point $\boldsymbol{q} \in \mathbb{R}^3$. In fact, an arbitrary point of this curve can be decomposed as

$$\boldsymbol{u} = \boldsymbol{u}_0\cos\alpha - \boldsymbol{u}_0^*\sin\alpha \tag{7.21}$$

where

$$\boldsymbol{u}_0 = \left(\sqrt{\frac{r - q_1}{2}}, \frac{q_2}{2\sqrt{2(r + q_1)}}, \frac{q_3}{2\sqrt{2(r + q_1)}}, 0\right) \tag{7.22}$$

is the point on the curve corresponding to $\alpha = 0$. Thus, the curve lies in the intersection of the plane spanned by $\boldsymbol{u}_0$ and $-\boldsymbol{u}_0^*$ and the sphere with center at the origin and radius $\sqrt{r}$. It follows that it is actually a circle, and it is easy to see that the tangent vector at the point $\boldsymbol{u}$ is given by $-\boldsymbol{u}^*$. •

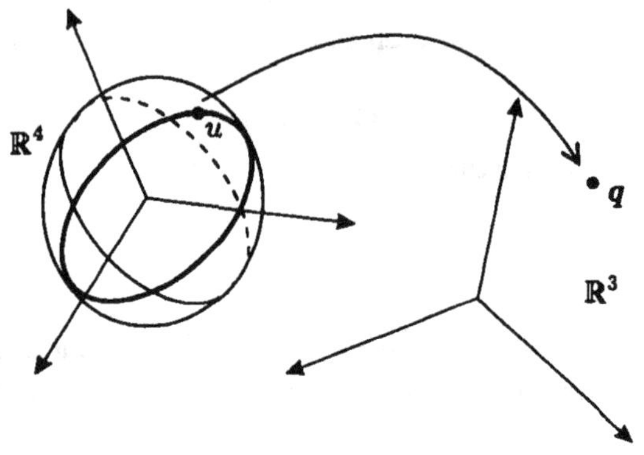

The KS transformation maps an entire circle in $\mathbb{R}^4$ onto one point in $\mathbb{R}^3$.

7.2 The photo-gravitational three-body problem

As a model of the physical situation described in Chapter 5 we consider the system depicted below. The asteroid and sun are here modeled with two primaries of mass M_1 and M_2 that orbit their common center of mass on circular orbits. Due to the vanishingly small mass of the satellite, its gravitational influence on the motion of the primaries is completely negligible. Its own dynamics, however, are highly dependent on the motion of the primaries. The model we have so far described is usually known as the **circular restricted three-body problem.** To account for the influence of radiation forces on the grain, we further include a pressure force field directed radially away from the star and satisfying Eq. (5.15).

In an inertial frame moving with the center of mass of the primaries, the motion is described by the Hamiltonian

$$H(\boldsymbol{q},\boldsymbol{p},q_0,p_0) = \frac{1}{2}|\boldsymbol{p}|^2 - \frac{GM_1}{r_1} + (1-\nu)GM_2\left(\frac{1}{R} - \frac{1}{r_2}\right) + p_0, \tag{7.23}$$

where we have introduced time and the energy as conjugate variables q_0, p_0 according to the suspended formalism. Here,

$$r_1^2 = (R_1\cos\omega q_0 - q_1)^2 + (R_1\sin\omega q_0 - q_2)^2 + q_3^2 \tag{7.24}$$

$$r_2^2 = (R_2\cos\omega q_0 + q_1)^2 + (R_2\sin\omega q_0 + q_2)^2 + q_3^2 \tag{7.25}$$

and, furthermore,

$$M_1R_1 = M_2R_2,\ R = R_1 + R_2,\ \text{and}\ \omega^2 = \frac{G(M_1+M_2)}{R^3}. \tag{7.26}$$

The constant term in the Hamiltonian has been included only to further simplify later expressions.

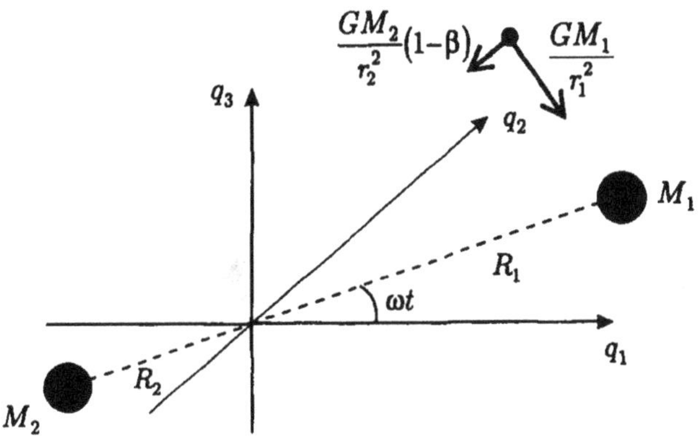

The circular restricted three-body problem. The motion of the primaries is independent of that of the grain.

We introduce the generating function

$$S_{rot}(\bar{q}_0, \bar{\boldsymbol{q}}, p_0, \boldsymbol{p}) = -\begin{pmatrix} p_0 & p_1 & p_2 & p_3 \end{pmatrix} \begin{pmatrix} \bar{q}_0 \\ (\bar{q}_1 + R_1)\cos\omega\bar{q}_0 - \bar{q}_2 \sin\omega\bar{q}_0 \\ (\bar{q}_1 + R_1)\sin\omega\bar{q}_0 + \bar{q}_2 \cos\omega\bar{q}_0 \\ \bar{q}_3 \end{pmatrix}, \quad (7.27)$$

which transforms the Hamiltonian into

$$H(\bar{\boldsymbol{q}}, \bar{\boldsymbol{p}}, \bar{q}_0, \bar{p}_0) = \frac{1}{2}|\bar{\boldsymbol{p}}|^2 - \frac{GM_1}{r_1} + (1-\nu)GM_2\left(\frac{1}{R} - \frac{1}{r_2}\right) + \bar{p}_0$$
$$+\omega(\bar{p}_1\bar{q}_2 - \bar{p}_2\bar{q}_1) - \omega R_1\bar{p}_2, \quad (7.28)$$

where $r_1 = |\bar{\boldsymbol{q}}|$ and $r_2^2 = r_1^2 + 2R\bar{q}_1 + R^2$. In the new coordinates, the asteroid lies permanently at the origin, while the star is located on the negative $\bar{q}_1$ axis. It is further convenient to consider the canonical transformation

$$S = \bar{q}_0(\tilde{p}_0 + \frac{1}{2}\omega^2 R_1^2) + \bar{q}_1\tilde{p}_1 + \bar{q}_2(\tilde{p}_2 + \omega R_1) + \bar{q}_3\tilde{p}_3. \quad (7.29)$$

The Hamiltonian now becomes

$$H = \frac{1}{2}|\tilde{\boldsymbol{p}}|^2 - \frac{GM_1}{r_1} + (1-\nu)GM_2\left(\frac{1}{R} - \frac{1}{r_2}\right) + \tilde{p}_0 + \omega(\tilde{p}_1\tilde{q}_2 - \tilde{p}_2\tilde{q}_1) - \omega^2 R_1\tilde{q}_1, \quad (7.30)$$

where $r_1 = |\tilde{\boldsymbol{q}}|$ and $r_2^2 = r_1^2 + 2R\tilde{q}_1 + R^2$. This can be further simplified by using Eq. (7.26) to write $\omega^2 R_1 = GM_2/R^2$.

Clearly, the Hamiltonian becomes singular as $r_1, r_2 \to 0$. However, since we are primarily concerned with motion around the asteroid, r_2 is safely bounded away

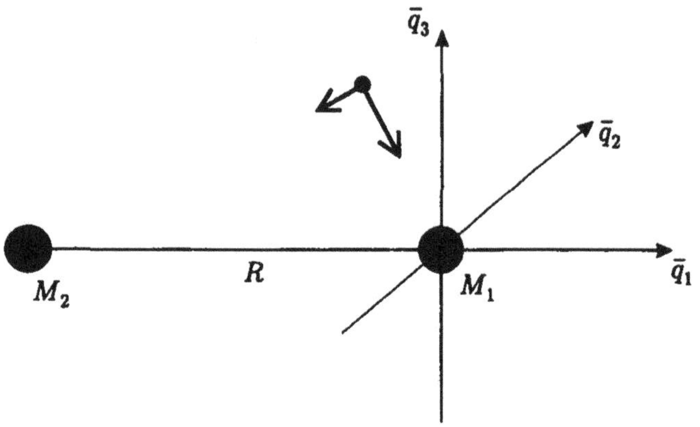

The perspective of an observer fixed relative to the asteroid-sun line.

from zero, while the same is not necessarily true of r_1. To eliminate the singularity, we thus introduce the time rescaling

$$\frac{dt}{ds} = r_1, \tag{7.31}$$

which results in the Hamiltonian

$$H = \frac{r_1}{2}|\tilde{\boldsymbol{p}}|^2 - GM_1 + GM_2 r_1 \left[(1-\nu)\left(\frac{1}{R} - \frac{1}{r_2}\right) - \frac{\tilde{q}_1}{R^2}\right]$$

$$+r_1\tilde{p}_0 + r_1\omega(\tilde{p}_1\tilde{q}_2 - \tilde{p}_2\tilde{q}_1). \tag{7.32}$$

We now apply the canonical KS transformation derived in the previous section. Using $l(\boldsymbol{w}, \boldsymbol{u}) = 0$ the transformed Hamiltonian can be expressed as

$$H_{KS}(u_0, \boldsymbol{u}, w_0, \boldsymbol{w}) = \frac{1}{8}|\boldsymbol{w}|^2 - GM_1 + w_0 r_1 + \frac{\omega}{2} r_1 (u_2 w_1 - u_1 w_2 - u_3 w_4 + u_4 w_3)$$

$$+Pr_1\left[\left(\frac{1}{\nu} - 1\right)\left(R - \frac{R^2}{r_2}\right) - \frac{u_1^2 - u_2^2 - u_3^2 + u_4^2}{\nu}\right], \tag{7.33}$$

where $r_1 = |\boldsymbol{u}|^2$ and $r_2^2 = r_1^2 + 2R(u_1^2 - u_2^2 - u_3^2 + u_4^2) + R^2$. We have suggestively introduced the parameter $P = \nu GM_2/R^2$, cf. Chapter 5.

We see that u_0 is cyclic, i.e., w_0 is conserved. As in Chapter 5, we reduce the number of physical parameters by the variable rescaling

$$\boldsymbol{u} = \sqrt{\frac{e}{2P}}\bar{\boldsymbol{u}}, \ \boldsymbol{w} = \frac{e}{2\sqrt{P}}\bar{\boldsymbol{w}}, \text{ and } s = \sqrt{\frac{2}{e}}\bar{s}, \tag{7.34}$$

where $e = |w_0|$. The evolution of the canonical coordinates $\bar{\boldsymbol{u}}$ and $\bar{\boldsymbol{w}}$ is now governed by the Hamiltonian

$$\bar{H}_{KS}(\bar{\boldsymbol{u}}, \bar{\boldsymbol{w}}) = \frac{1}{8}|\bar{\boldsymbol{w}}|^2 + 2\mathrm{sign}(w_0)\bar{r}_1 + \bar{r}_1\left[\left(\frac{1}{\nu} - 1\right)\left(\bar{R} - \frac{\bar{R}^2}{\bar{r}_2}\right) + \frac{\bar{u}_1^2 - \bar{u}_2^2 - \bar{u}_3^2 + \bar{u}_4^2}{\nu}\right]$$

$$-\frac{4PGM_1}{e^2} + \varepsilon\bar{r}_1(\bar{u}_2\bar{w}_1 - \bar{u}_1\bar{w}_2 - \bar{u}_3\bar{w}_4 + \bar{u}_4\bar{w}_3), \qquad (7.35)$$

where $\bar{r}_1 = |\bar{\boldsymbol{u}}|^2$, $\bar{r}_2^2 = \bar{r}_1^2 + 2\bar{R}(\bar{u}_1^2 - \bar{u}_2^2 - \bar{u}_3^2 + \bar{u}_4^2) + \bar{R}^2$, $\bar{R} = \frac{2P}{e}R$, and

$$\varepsilon = \frac{\sqrt{e}}{2P\sqrt{2}}\omega. \qquad (7.36)$$

We note that it is still true that $\bar{H}_{KS} \equiv 0$.

We proceed to introduce polar coordinates through the generating function

$$S_{polar}(\tilde{\boldsymbol{u}}, \bar{\boldsymbol{w}}) = -\begin{pmatrix} \bar{w}_1 & \bar{w}_2 & \bar{w}_3 & \bar{w}_4 \end{pmatrix}\begin{pmatrix} \tilde{u}_1\cos\tilde{u}_3 \\ \tilde{u}_2\cos\tilde{u}_4 \\ \tilde{u}_2\sin\tilde{u}_4 \\ \tilde{u}_1\sin\tilde{u}_3 \end{pmatrix} \qquad (7.37)$$

which yields

$$H_{polar}(\tilde{\boldsymbol{u}}, \tilde{\boldsymbol{w}}) = \frac{1}{8}\left(\tilde{w}_1^2 + \tilde{w}_2^2 + \frac{\tilde{w}_3^2}{\tilde{u}_1^2} + \frac{\tilde{w}_4^2}{\tilde{u}_2^2}\right) + 2\mathrm{sign}(w_0)\bar{r}_1 - \frac{4PGM_1}{e^2}$$

$$+\bar{r}_1\left[\left(\frac{1}{\nu} - 1\right)\left(\bar{R} - \frac{\bar{R}^2}{\bar{r}_2}\right) - \frac{\tilde{u}_1^2 - \tilde{u}_2^2}{\nu}\right]$$

$$+\varepsilon\bar{r}_1\left[(\tilde{u}_2\tilde{w}_1 - \tilde{u}_1\tilde{w}_2)\cos(\tilde{u}_3 + \tilde{u}_4) + \left(\frac{\tilde{u}_1\tilde{w}_4}{\tilde{u}_2} - \frac{\tilde{u}_2\tilde{w}_3}{\tilde{u}_1}\right)\sin(\tilde{u}_3 + \tilde{u}_4)\right], \qquad (7.38)$$

where $\bar{r}_1 = \tilde{u}_1^2 + \tilde{u}_2^2$, and $\bar{r}_2^2 = \bar{r}_1^2 + 2\bar{R}(\tilde{u}_1^2 - \tilde{u}_2^2) + \bar{R}^2$. In the tilded coordinates the submanifold $l(\boldsymbol{w}, \boldsymbol{u}) = 0$ corresponds to $\tilde{w}_3 = \tilde{w}_4$. We further note that the angle-type position coordinates $\tilde{u}_3$ and $\tilde{u}_4$ only appear in the combination $\tilde{u}_3 + \tilde{u}_4$. Thus, an equivalent system is obtained by letting $\tilde{w}_4 = \tilde{w}_3 = J$ and $\tilde{u}_3 + \tilde{u}_4 = \theta$. This is easily seen by introducing the canonical transformation $\tilde{u}_4 \to \tilde{u}_4 - \tilde{u}_3$ and $\tilde{w}_3 \to \tilde{w}_3 + \tilde{w}_4$ and observing that $\tilde{u}_3$ is cyclic in the new Hamiltonian.

Finally, we assume that $\bar{R} \gg \tilde{u}_i^2$, which allows us to expand the $\frac{1}{\bar{r}_2}$ term in inverse powers of $\bar{R}$:

$$\frac{1}{\bar{r}_2} = \frac{1}{\bar{R}} - \frac{\tilde{u}_1^2 - \tilde{u}_2^2}{\bar{R}^2} + \mathcal{O}(\frac{1}{\bar{R}^3}). \qquad (7.39)$$

Omitting higher order terms, our final expression for the Hamiltonian is

$$H(\tilde{u}_1,\tilde{u}_2,\theta,\tilde{w}_1,\tilde{w}_2,J)=\frac{1}{8}\left(\tilde{w}_1^2+\tilde{w}_2^2+\frac{J^2}{\tilde{u}_1^2}+\frac{J^2}{\tilde{u}_2^2}\right)+2\text{sign}(w_0)(\tilde{u}_1^2+\tilde{u}_2^2)-\tilde{u}_1^4+\tilde{u}_2^4$$

$$+\varepsilon(\tilde{u}_1^2+\tilde{u}_2^2)\left[(\tilde{u}_2\tilde{w}_1-\tilde{u}_1\tilde{w}_2)\cos\theta+J\left(\frac{\tilde{u}_1}{\tilde{u}_2}-\frac{\tilde{u}_2}{\tilde{u}_1}\right)\sin\theta\right] \tag{7.40}$$

where $H\equiv 4PGM_1/e^2$. Guided by the analysis in Chapter 5, we will take $w_0>0$, since otherwise all trajectories are unbounded.

Illustration

In the expression above we omitted terms of order

$$\mathcal{O}((\frac{1}{\nu}-1)\frac{1}{\bar{R}^n}),\ \textit{for}\ n=1,2,\ldots \tag{7.41}$$

Using Eq. (7.36), the definition for P, and Eq. (7.26), it is easy to show that

$$\frac{\varepsilon^{2n}}{(1/\bar{R}^n)}=(4\nu)^{-n}\left(\frac{M_1+M_2}{M_2}\right)^n\approx(4\nu)^{-n}$$

where we have used the fact that $M_2\gg M_1$. *For small values of* ν, *the omitted terms are thus of order*

$$\mathcal{O}((4\nu)^{n-1}\varepsilon^{2n}),\ \textit{for}\ n=1,2,\ldots \tag{7.42}$$

Thus, for small ε *these terms can be neglected in a lower-order perturbation analysis. However, since* $\varepsilon\sim\sqrt{R}$, *the assumption that* $\varepsilon\ll 1$ *is likely to fail at greater distances from the sun. Finally, we note from Eq. (7.36) that the value of* ε *depends on the value of the first integral* e. *Thus, a perturbation approach might only be valid in certain regions of the original state space.* •

7.3 In the absence of rotation

When $\varepsilon=0$, the Hamiltonian system given by Eq. (7.40) is completely integrable. In particular, since θ is cyclic for the unperturbed Hamiltonian, J is a conserved quantity and functions as a parameter for the remaining dynamics. Moreover, for positive J, θ increases with an angular velocity dependent on the dynamics in the $(\tilde{u}_1,\tilde{w}_1)$ and $(\tilde{u}_2,\tilde{w}_2)$ degrees-of-freedom. Finally, since the unperturbed Hamiltonian decouples the $(\tilde{u}_1,\tilde{w}_1)$ and $(\tilde{u}_2,\tilde{w}_2)$ systems, we are left with analyzing two one-degree-of-freedom Hamiltonians, with one free parameter. In particular, we introduce

$$F_1(\tilde{u}_1,\tilde{w}_1,J)=\frac{1}{8}\tilde{w}_1^2+2\tilde{u}_1^2-\tilde{u}_1^4+\frac{J^2}{8\tilde{u}_1^2} \tag{7.43}$$

and

$$F_2(\tilde{u}_2, \tilde{w}_2, J) = \frac{1}{8}\tilde{w}_2^2 + 2\tilde{u}_2^2 + \tilde{u}_2^4 + \frac{J^2}{8\tilde{u}_2^2}. \tag{7.44}$$

Illustration

We consider the types of motion available in the $(\tilde{u}_1, \tilde{w}_1)$ system. In particular, the Hamiltonian is of the form $H(p,q) = \frac{1}{8}p^2 + V(q)$, from which it follows that motion is only possible for which $H \geq V(q)$. Thus, the qualitative dynamics are determined by the shape of V, the potential. In our case, we thus study the function

$$V(q) = 2q^2 - q^4 + \frac{J^2}{8q^2}. \tag{7.45}$$

for $q > 0$. To determine the existence of extrema of the potential, we study the zeros of $V'(q)$, or equivalently, of the function

$$g(q) = 16q^4 - 16q^6 - J^2.$$

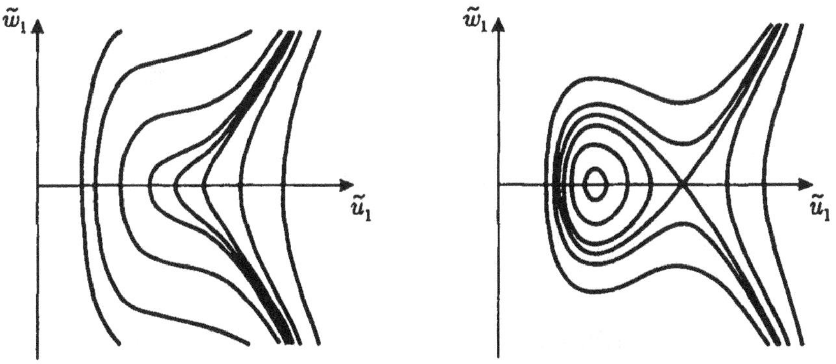

Phase diagrams for $J^2 > \frac{64}{27}$ (left panel) and $J^2 < \frac{64}{27}$ (right panel).

It is easy to show that two extrema exist when $J^2 \in \left(0, \frac{64}{27}\right)$, while there are no extrema when $J^2 > \frac{64}{27}$. Since $V \to \infty$ as $q \to 0$, and $V \to -\infty$ when $q \to \infty$, this information allows us to draw the state-space diagrams shown above. Clearly, the existence of bounded motion in the $(\tilde{u}_1, \tilde{w}_1)$ system is limited to the case in which $J^2 \in \left(0, \frac{64}{27}\right)$. For $J^2 > \frac{64}{27}$ all orbits are unbounded. In the limiting case in which $J^2 = \frac{64}{27}$, the only bounded motion is stationary at the saddle of the potential. •

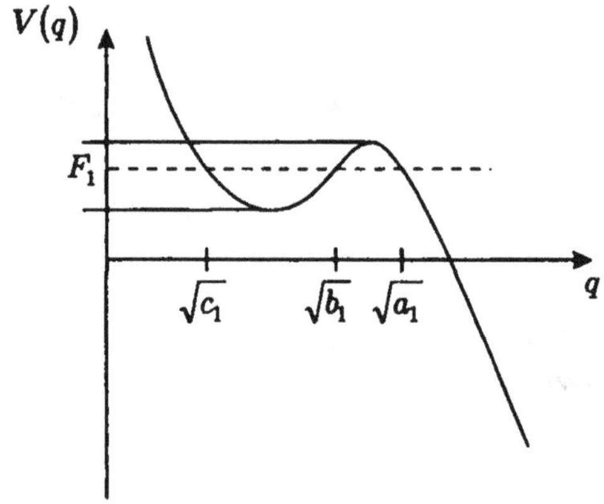

The potential when $J^2 < \frac{64}{27}$.

For values of F_1 in the interval shown in the figure above, there are three values of $\tilde{u}_1$ for which $\tilde{w}_1 = 0$. Denote these by $\sqrt{a_1} \geq \sqrt{b_1} > \sqrt{c_1}$. From Eq. (7.43) it follows that a_1, b_1 and c_1 satisfy the cubic

$$t^3 - 2t^2 + F_1 t - \frac{J^2}{8} = 0. \tag{7.46}$$

From the expressions for the coefficients of the cubic in terms of its roots, we obtain the system of equations

$$c_1 + b_1 + a_1 = 2, \tag{7.47}$$

$$a_1 b_1 + b_1 c_1 + a_1 c_1 = F_1, \tag{7.48}$$

$$a_1 b_1 c_1 = \frac{J^2}{8}. \tag{7.49}$$

We are particular interested in the case when $a_1 = b_1$, corresponding to motion on the homoclinic connection separating bounded from unbounded motion for which $F_1 = F_{sep}$. Then, Eqs. (7.47-7.49) can actually be solved to yield

$$b_1 = \frac{2 + \sqrt{4 - 3F_{sep}}}{3} = b^*,\ c_1 = \frac{2 - 2\sqrt{4 - 3F_{sep}}}{3} = c^*, \tag{7.50}$$

where

$$\frac{27}{16} J^2 = 4 - 3\left(4 - 3F_{sep}\right) - \left(4 - 3F_{sep}\right)^{3/2}. \tag{7.51}$$

It follows that the separatrix is homoclinic to a saddle point at

$$\left(\tilde{u}_1^*, \tilde{w}_1^*\right) = \left(\sqrt{b^*}, 0\right), \tag{7.52}$$

where F_{sep} can be shown to be the unique solution in $(1, \frac{4}{3})$ to Eq. (7.51). The separatrix trajectory can further be solved for explicitly as a function of time:

$$\tilde{u}_{1,0}^2(s) = c^* + (b^* - c^*) \tanh^2 s \sqrt{\frac{b^* - c^*}{2}} \tag{7.53}$$

and

$$\tilde{w}_{1,0}(s) = 4 \frac{d}{ds} \tilde{u}_{1,0}(s). \tag{7.54}$$

As advertised in previous chapters, the perturbation analysis is primarily interested in motion near the unperturbed separatrix. For our purposes it suffices to note that the period of unperturbed, bounded oscillatory solutions close to the separatrix is given by

$$T = \frac{2\sqrt{2}}{\sqrt{a_1 - c_1}} \mathrm{K}(k_1) \approx \frac{2\sqrt{2}}{\sqrt{a_1 - c_1}} \ln \frac{4}{k_1'}, \text{ for } k_1' \ll 1 \tag{7.55}$$

where $k_1' = \sqrt{\frac{a_1 - b_1}{a_1 - c_1}}$, and K denotes the complete elliptic integral of the first kind.

For the $(\tilde{u}_2, \tilde{w}_2)$ system one may similarly show that for all values of J all motions are oscillatory. In particular, $F_2 > F_{min}$, where F_{min} is the unique solution in $(0, \frac{5}{3})$ to

$$(4 + 3x)^{\frac{3}{2}} - 9x - \frac{27}{16} J^2 - 8 = 0 \tag{7.56}$$

provided $J^2 \in (0, \frac{64}{27})$, cf. Eq. (7.51). To accommodate the discussion in the previous chapter, we introduce action-angle variables. Standard calculations yield the necessary coordinate transformation $(\tilde{u}_2, \tilde{w}_2, \theta, J) \to (\varphi_1, I_1, \varphi_2, I_2)$:

$$\tilde{u}_2^2 = c_2 + (b_2 - c_2) \mathrm{dn}^{-2} \left[\frac{\mathrm{K}(k_2)\varphi_1}{\pi}, k_2 \right], \; \tilde{w}_2 = 4 \frac{\partial \alpha}{\partial I_1} \frac{\mathrm{d}\tilde{u}_2}{\mathrm{d}\varphi_1}, \; J = I_2, \tag{7.57}$$

and

$$\theta = \varphi_2 - \frac{\partial \alpha}{\partial I_2} \left[\frac{\partial \alpha}{\partial I_1} \right]^{-1} \varphi_1 + \frac{I_2 g}{4\sqrt{2} b_2} \int_0^{\frac{\mathrm{K}(k_2)\varphi_1}{\pi}} \frac{\mathrm{dn}^2[u, k_2]}{1 - \frac{c_2}{b_2} k_2^2 \mathrm{sn}^2[u, k_2]} \mathrm{d}u. \tag{7.58}$$

Here, $\alpha(I_1, I_2) = F_2$, from which we obtain

$$\frac{\partial \alpha}{\partial I_1} = \frac{\pi \sqrt{2}}{g \mathrm{K}(k_2)} \tag{7.59}$$

and

$$\frac{\partial \alpha}{\partial I_2} = \frac{I_2}{4 b_2 \mathrm{K}(k_2)} \int_0^{\mathrm{K}(k_2)} \frac{\mathrm{dn}^2[u, k_2]}{1 - \frac{c_2}{b_2} k_2^2 \mathrm{sn}^2[u, k_2]} \mathrm{d}u. \tag{7.60}$$

Moreover,

$$g = \frac{2}{\sqrt{a_2 - c_2}} \text{ and } k_2 = \sqrt{\frac{a_2 - b_2}{a_2 - c_2}}. \tag{7.61}$$

Here, $a_2 > b_2 > c_2$ are the solutions to

$$x^3 + 2x^2 - F_2 x + \frac{J^2}{8} = 0. \tag{7.62}$$

Illustration

The unperturbed Hamiltonian can now be written

$$H_0(\tilde{u}_1, \tilde{w}_1, \boldsymbol{I}) = \frac{1}{8}\tilde{w}_1^2 + 2\tilde{u}_1^2 - \tilde{u}_1^4 + \frac{I_2^2}{8\tilde{u}_1^2} + \alpha(I_1, I_2), \tag{7.63}$$

where $(\tilde{u}_1, \tilde{w}_1, \varphi, \boldsymbol{I}) \in \mathbb{R}^2 \times \mathbb{T}^2 \times \mathbb{R}_+^2$. *We denote the pair* $(\tilde{u}_1, \tilde{w}_1)$ *by* $\boldsymbol{z}$. *From the above analysis it follows that for*

$$\boldsymbol{I} \in U = \left\{ \boldsymbol{I} \mid I_1 \in \mathcal{R}_+, I_2^2 \in \left(0, \frac{64}{27}\right) \right\},$$

the unperturbed $\boldsymbol{z}$ *degree-of-freedom contains a hyperbolic equilibrium point whose coordinates depend smoothly on* $\boldsymbol{I}$. *Moreover, the equilibrium point is connected to itself through a one-dimensional homoclinic manifold.*

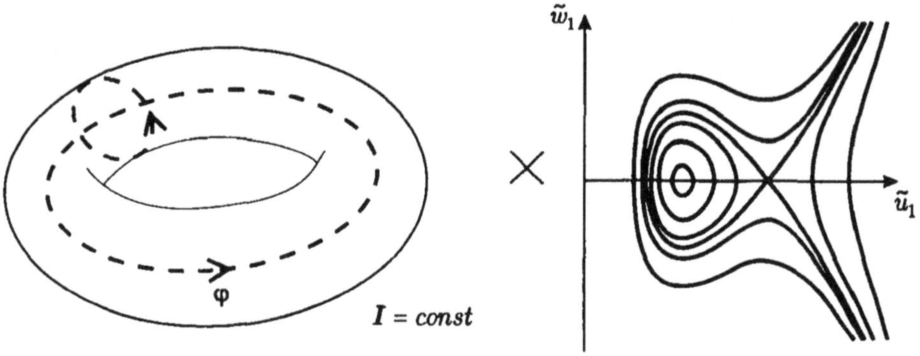

The full unperturbed dynamics decouple into tori cross motion in a single degree-of-freedom containing a homoclinic orbit.

As described in the previous chapter, the unperturbed state space restricted to the manifold of hyperbolic equilibria is foliated by two-dimensional, invariant tori, $\tau_0(\boldsymbol{I})$. *Moreover, the toral motion is quasiperiodic with frequencies given by the components of the vector*

$$\boldsymbol{D}_{\boldsymbol{I}} H_0(\tau_0(\boldsymbol{I}), \boldsymbol{I}).$$

In particular, the evolution in the $(\varphi, \boldsymbol{I})$ degrees-of-freedom as the $\boldsymbol{z}$ state moves along the homoclinic orbit is given by

$$\varphi_{1,0}(s) = \frac{\partial \alpha}{\partial I_1} s + \varphi_{1,0}(0), \tag{7.64}$$

$$\varphi_{2,0}(s) = \frac{\partial \alpha}{\partial I_2} s$$

$$+\frac{I_2}{b^*}\left\{\sqrt{\frac{2}{c^*}} \arctan\left[\sqrt{\frac{b^*-c^*}{c^*}} \tanh s\sqrt{\frac{b^*-c^*}{2}}\right] + s\right\} + \varphi_{2,0}(0), \tag{7.65}$$

and $I_i(s) \equiv I_i$ =constant, $i = 1, 2$. •

Each of the unperturbed two-dimensional tori is connected to itself through a three-dimensional homoclinic manifold. As in the previous chapter, we proceed to study the consequences of the perturbation on the tori and their corresponding manifolds.

7.4 In the presence of rotation

In the previous chapter we argued that under certain conditions on the frequency vector, $\boldsymbol{D_I} H_0(\tau_0(\boldsymbol{I}), \boldsymbol{I})$, the unperturbed invariant tori persist under sufficiently small perturbations. Similarly, the persistent tori, $\tau_\varepsilon(\boldsymbol{I})$, can be shown to possess three-dimensional perturbed stable and unstable manifolds $\mathcal{W}^{s,u}(\tau_\varepsilon(\boldsymbol{I}))$ on which orbits approach the perturbed torus as $t \to \infty$ and $-\infty$, respectively. Contrary to the integrable case, the perturbed manifolds are not expected to coincide. Instead they split, generally leaving behind points of intersection.

We further indicated the possibility of transitions between tori at large separation, through the process of Arnold diffusion, provided the stable and unstable manifolds of different persistent tori could be found sufficiently close to each other. These transition chains were expected to decrease in length under increasing perturbation strength, primarily due to the increasing separation between persistent tori. It was finally noted that the relevant time scales for these types of state-space motions were polynomially slow in the perturbation parameter.

In the present application, Arnold diffusion would enable a grain to sample regions in state space with qualitatively different dynamics. The above analysis indicated the presence of bounded as well as unbounded orbits in the unperturbed state space. In particular, for certain values of the action variables, all motion is unbounded. Of course, in the absence of rotation, the complete integrability of the Hamiltonian prevents transitions between these different types of behavior. The transversal intersections of the manifolds removes this restriction. As a result, large changes in the action variables may occur, eventually leading to a region of state space containing only unbounded motions. Similarly, the diffusion process allows particles initially bound to the asteroids to eventually escape, given sufficiently long time. It would therefore be valuable to estimate the diffusion time scale, and, in particular, to compare it to the age of the solar system, or relevant reinjection rates. This comparison would be an indicator of whether the diffusion process would be sufficiently fast to deplete neighborhoods of asteroids initially filled with small grains.

A rigorous application of the results of the previous chapter would entail proving that the gaps between the persistent tori are sufficiently small to allow transition chains to form. It was argued that this could be guaranteed, provided that low-order resonances were avoided and that the perturbation was sufficiently small. Rather than perform the necessary calculations, we attempt below to obtain estimates for possible diffusion rates, albeit with the understanding that their applicability might be limited to certain regions of state space.

Finally, we observe that the escape mechanism discussed here only applies to orbits originally close to the unperturbed separatrix. In fact, inside the separatrix, standard KAM theory implies the persistence of a majority of nonresonant invariant tori, i.e., orbits bounded for all times. In addition to these, there is generically a complementary, open set of points contained in the stochastic layers. As argued in Chapter 2, the high-dimensionality of state space would allow these points to drift out toward the unperturbed separatrix, where the mechanism discussed above might apply. Of course, even without changes in the action variables, orbits may escape by essentially crossing the unperturbed separatrix. This would only add to the diffusion rate, and so estimating the escape rate due to Arnold diffusion along the transition chain at least provides a reasonable lower bound.

In the sections below, we apply the theory of the previous chapter to prove the existence of transversal intersections between the stable and unstable manifolds of persistent tori. Moreover, these results are used to obtain an estimate for the escape rate which provides a relevant starting point in a physical comparison with rates connected to other processes.

7.5 The manifold separation

In the previous chapter we showed how the manifold separation to first order was determined by studying the indefinite integrals

$$\int_0^t \frac{\partial H_1}{\partial \varphi}(\varphi_0(s), \boldsymbol{I}_0, z_0(s))ds, \tag{7.66}$$

in the limit when $t \to \pm\infty$. In particular, the two-dimensional first-order separation vector was given by

$$\varepsilon \Delta \boldsymbol{\delta} = \varepsilon \left(\boldsymbol{\delta}^s - \boldsymbol{\delta}^u\right), \tag{7.67}$$

where $\boldsymbol{\delta}^{s,u}$ denote the constant part of the quasiperiodic function approached by the integral (7.66) as $t \to \pm\infty$. Clearly, a vanishing separation vector for particular values of $\varphi_0(0)$ implies the existence of points of intersection of the perturbed manifolds to first order in ε.

We require, however, that the manifolds intersect to all orders in ε, i.e., that the first-order result persists under small perturbations. In the previous chapter, this was guaranteed by a transversality condition. In particular, the simplest such condition was given by the nonvanishing of the determinant

$$\left|\frac{\partial \Delta \delta}{\partial \varphi_0(0)}\right|. \tag{7.68}$$

Illustration

Consider the system of equations

$$h(x, y, \varepsilon) = 0 \tag{7.69}$$

$$g(x, y, \varepsilon) = 0 \tag{7.70}$$

such that $(0,0,0)$ *is a solution and* $g_y(0,0,0) \neq 0$. *Then, by the implicit function theorem, the latter condition allows us to solve Eq. (7.70) uniquely for* $y = y(x, \varepsilon)$ *in a suitable neighborhood of the origin. Substituting this result back into Eq. (7.69) yields the equation*

$$f(x, \varepsilon) = h(x, y(x, \varepsilon), \varepsilon) = 0, \tag{7.71}$$

which has a solution at $(0,0)$. *Indeed, the solution will persist for small deviations in* ε *from* 0, *provided that the zero of the function* $f(x, 0)$ *at the origin is of odd order. Of course, unless the zero is simple, the solution may no longer be unique, but this is not necessary for our purposes.*

From Eq. (7.71) it follows that the order of the zero at the origin can be determined by differentiation of h *and* g *with respect to* x *and* y. *In particular, we find*

$$f_x(x, \varepsilon) = h_x(x, y(x, \varepsilon), \varepsilon) + h_y(x, y(x, \varepsilon), \varepsilon)\, y_x(x, \varepsilon). \tag{7.72}$$

But,

$$g_x(x, y(x, \varepsilon), \varepsilon) + g_y(x, y(x, \varepsilon), \varepsilon)\, y_x(x, \varepsilon) = 0 \tag{7.73}$$

implies that

$$y_x(x, \varepsilon) = -\frac{g_x(x, y(x, \varepsilon), \varepsilon)}{g_y(x, y(x, \varepsilon), \varepsilon)}, \tag{7.74}$$

since $g_y \neq 0$ *in a neighborhood of the origin. Consequently,*

$$f_x(x, \varepsilon) = h_x(x, y(x, \varepsilon), \varepsilon) - h_y(x, y(x, \varepsilon), \varepsilon)\frac{g_x(x, y(x, \varepsilon), \varepsilon)}{g_y(x, y(x, \varepsilon), \varepsilon)}, \tag{7.75}$$

and the origin is a simple zero if

$$\left|\frac{\partial(h,g)}{\partial(x,y)}\right|(0,0,0)\neq 0, \tag{7.76}$$

which in the present case corresponds to the transversality condition (7.68).

In actual applications the conditions can generally only be checked numerically. To further support one's conclusions, it is therefore useful to check higher-order transversality conditions. In the radiation pressure problem considered here, it can be shown that, by symmetry, the even-order derivatives of f in the illustration vanish. The next higher transversality condition is given by considering the third derivative of f:

$$f_{xxx} = h_{xxx} + 3h_{xxy}y_x + 3h_{xyy}y_x^2 + h_{yyy}y_x^3 + h_y y_{xxx}, \tag{7.77}$$

where

$$y_{xxx} = -\left(g_{xxx} + 3g_{xxy}y_x + 3g_{xyy}y_x^2 + g_{yyy}y_x^3\right)/g_y. \tag{7.78}$$

•

We now apply the theory to our Hamiltonian (7.40) where

$$H_1(\tilde{u}_1,\tilde{w}_1,\varphi,I)$$

$$= \left(\tilde{u}_1^2+\tilde{u}_2^2\right)\left[\left(\tilde{u}_2\tilde{w}_1-\tilde{u}_1\tilde{w}_2\right)\cos\theta + I_2\left(\frac{\tilde{u}_1}{\tilde{u}_2}-\frac{\tilde{u}_2}{\tilde{u}_1}\right)\sin\theta\right], \tag{7.79}$$

and $\tilde{u}_2$, $\tilde{w}_2$, and θ are given in Eqs. (7.57-7.58). In order to find the constant parts of the integral $\int_0^t \frac{\partial H_1}{\partial\varphi}dt'$ as $t\to\pm\infty$, we expand H_1 in a complex Fourier series in the angles φ

$$\sum_{m,n} a_{m,n}\left(\tilde{u}_1,\tilde{w}_1\right)e^{i(m\varphi_1+n\varphi_2)}, \tag{7.80}$$

where in our case the expression for θ implies that $a_{m,n}=0$ for $n\neq\pm1$.

Since H_1 is analytic the Fourier series can be differentiated and integrated term by term. Moreover, derivatives with respect to φ and $\varphi_0(0)$ appear simply as additional factors of n and/or m while the involved integrals are unchanged. It thus suffices to consider integrals of the general form

$$\int_0^t f(s)e^{i\psi(s)}ds, \tag{7.81}$$

where $\psi = m\varphi_1 + n\varphi_2$, and f is expressed in terms of the homoclinic solution (7.53-7.54).

In particular, $f \to$ constant exponentially fast as $t \to \pm\infty$. If the limiting constant is zero, then the integral above is absolutely convergent as $t \to \pm\infty$ and the constant part of the integral is simply given by its value as $t \to \pm\infty$. If, instead, f approaches a nonzero constant, then, assuming that $\dot{\psi}(s) \neq 0$, $s \in [0, t]$, we integrate by parts to obtain

$$\int_0^t f(s)e^{i\psi(s)}ds = \left[\frac{f(s)}{i\dot{\psi}(s)}e^{i\psi(s)}\right]_0^t - \frac{1}{i}\int_0^t \frac{\dot{f}(s)\dot{\psi}(s) - f(s)\ddot{\psi}(s)}{[\dot{\psi}(s)]^2}e^{i\psi(s)}ds. \tag{7.82}$$

Since $\dot{\psi}$ does not vanish identically, the upper limit in the first term on the right hand side does not contribute any constant term to the quasiperiodic function. The lower limit, on the other hand, is identical for $t \to \pm\infty$ and thus vanishes when the difference $\Delta\delta$ is computed. The remaining integral is absolutely convergent as $t \to \pm\infty$ and can consequently be straightforwardly evaluated numerically.

It remains to consider the case when $\dot{\psi} = 0$ at some time. Since $\dot{\psi}$ depends on the motion along the homoclinic orbit, it is generally possible that one or several resonances between the unperturbed frequencies will be encountered. If the perturbation contains harmonics of order (m, n) for all $m, n \in Z$, then $\dot{\psi}$ at a given time can be made arbitrarily small for sufficiently large values of m and n. However, since we are only concerned with the tori which survive the perturbation, and since the motion on these is nonresonant, then for every pair (m, n) there is a time T such that $\dot{\psi} \neq 0$ for $t > T$. Hence, the evaluation can be divided into two steps. The first is the calculation of the definite integral over $[0, T]$ and the second is using the expression (7.82) for which the integral is absolutely convergent. Since, in our case, $n = \pm 1$, only finitely many resonances will be encountered. In particular, there are choices for $\boldsymbol{I}$ such that no resonances are encountered and thus Eq. (7.82) applies unmodified.

The implementation of the above considerations in our problem leads to highly non-trivial integrands, for which only numerical methods are available. Because of possible numerical errors it is preferable to use symmetries for identifying values of $\varphi_0(0)$ for which $\Delta\delta$ vanishes. In our case, it is easily shown that the odd/even symmetries in the integrands imply the vanishing of $\Delta\delta$ for $\varphi_0(0) = (0, \frac{\pi}{2})$ regardless of $\boldsymbol{I}$. As we previously hinted, the same symmetry implies the vanishing of all even-order transversality conditions. If we can now show that a transversality condition is satisfied for a particular value of $\boldsymbol{I} = \tilde{\boldsymbol{I}}$, then by continuity it is satisfied for values $\boldsymbol{I}$ in a neighborhood of $\tilde{\boldsymbol{I}}$. From the denseness of the perturbed tori as $\varepsilon \to 0$ it follows that the transversality condition is satisfied for a number of perturbed tori. This in turn implies the transversal intersection of the perturbed manifolds of these tori and, for sufficiently small ε, the intersection of manifolds belonging to different tori, thus setting up the mechanism for motions along the transition chain, as discussed previously.

In checking the transversality conditions, the accuracy of the numerical evaluations becomes of crucial importance, when differentiating between zero and nonzero quantities. In particular, we showed in Chapter 5 that the first-order manifold separation vanished identically in a two-degree-of-freedom problem. Indeed, to order ε, that calculation corresponds in the problem of this chapter to restricting the motion to the plane of the primaries. It is not expected that the manifolds continue to coincide to $\mathcal{O}(\varepsilon)$ in the full three-degree-of-freedom case. However, for state-space motions near the invariant two-degree-of-freedom manifold, the splitting of the manifolds might be difficult to discern.

For $F_{sep} = \frac{5}{4}$ and $F_2 = 5$ a value of $\approx 1.5 \cdot 10^{-5}$ is obtained for the linear transversality condition and $\approx 6 \cdot 10^{-4}$ for the cubic condition. While both these numbers are small, they show the transversality of the manifold intersection. In light of the remarks above we point out that, for certain values of F_{sep} and F_2, the determinant (7.68) is on the order of $10^{-8} \sim 10^{-10}$, at which point the first-order analysis becomes questionable.

7.6 Arnold diffusion

We proceed to estimate the diffusion rate discussed above. We will attempt to estimate the change in action $\boldsymbol{I}$ during one period of a perturbed motion near the original homoclinic manifold. An appropriate measure for the period of such a motion is obtained by averaging T from Eq. (7.55) over an interval $[0, k']$ in the complementary modulus yielding

$$T_{av} \approx \frac{1}{k_1'} \int_0^{k_1'} \frac{2\sqrt{2}}{\sqrt{a_1 - c_1}} \ln \frac{4}{s} ds \approx \frac{2\sqrt{2}}{\sqrt{a_1 - c_1}} \left(\ln \frac{4}{k_1'} + 1 \right), \tag{7.83}$$

where to a first approximation $a_1 - c_1$ is constant.

It remains to obtain an appropriate value for k_1'. Before we do this, we show how the change in action can be evaluated from the discussion of the previous sections. As shown in Chapter 6, to first order, the change in action for a given perturbed orbit is given by $\varepsilon\Delta\boldsymbol{\delta}$. We take the root-mean-square value of each of the two components averaged over all values of $\varphi_0(0)$ as a first-order estimate of the change in $\boldsymbol{I}$ per period of oscillation.

The first-order distance between the perturbed manifolds in the $(\tilde{u}_1, \tilde{w}_1)$ plane, $\varepsilon\Delta\beta$, is given from the $\Delta\boldsymbol{\delta}$ through the expression

$$\Delta\beta = -\frac{\partial H_0}{\partial \boldsymbol{I}} \cdot \Delta\boldsymbol{\delta}, \tag{7.84}$$

where the partial derivatives are evaluated on the unperturbed torus, cf. Eq. (6.77). Again, the root mean square value of $\varepsilon\Delta\beta$ provides a measure for the thickness of the separatrix layer near the original homoclinic connection and can thus be used to estimate a value for k_1' in Eq. (7.83). For our purposes, we let $b_1 = b^* - \varepsilon\sqrt{\langle|\Delta\beta|^2\rangle}$ and solve the necessary equations from Section 3 for k_1'.

We apply the discussion of the previous paragraphs to the case $F_2 = 5$ and $F_{sep} = \frac{5}{4}$, for which we previously showed existence of transversal intersections of the perturbed manifolds. The numerical calculations then yield

$$\sqrt{\langle|\Delta\delta_1|^2\rangle} \approx 0.0706, \ \sqrt{\langle|\Delta\delta_2|^2\rangle} \approx 0.0706, \ \sqrt{\langle|\Delta\beta|^2\rangle} \approx 0.000133$$

To proceed we need an estimate for the small parameter ε, which in turn requires estimating the value of the conserved quantity w_0 as shown in Eq. (7.36). Recall Eq. (7.40):

$$\frac{4GM_1P}{e^2} = F_1 + F_2 + \varepsilon H_1. \tag{7.85}$$

For simplicity choose initial conditions such that at $s = 0$, $H_1 = 0$ and $F_1 = F_{sep}$. This implies that

$$e = \frac{4}{5}\sqrt{GM_1P} \tag{7.86}$$

and furthermore

$$\varepsilon = \frac{\omega}{2P}\sqrt{\frac{2}{5}\sqrt{GM_1P}}. \tag{7.87}$$

We consider a small grain of millimeter size in orbit around the asteroid Eros which is due to be visited by the NEAR satellite in 1998. Eros has an estimated mass of 5×10^{15} kg, an approximate orbital period of 1.8 yrs and lies at a distance of about 1.5 au from the sun. In Chapter 5 we derived for the dimensionless quantity ν the expression

$$\nu = 5.7 \times 10^{-7}\frac{Q}{R\rho}\ \mathrm{kgm}^{-2}, \tag{7.88}$$

where it is assumed that the peak wavelength of the solar radiation is much smaller than the particle size. Here, R is the radius of the grain, ρ is the grain density, and $Q \approx 1$ is determined from the optical properties of the grain. Assuming a density of ≈ 2 g/cm^3, the above values imply $P \approx 8 \times 10^{-7}$ m/s^2. From the orbital period of Eros we obtain $\omega \approx 1 \times 10^{-7}$ s^{-1}. Substitution into the above relations then yields $e \approx 0.4$ and $\varepsilon \approx 0.03$. The smallness of this parameter makes it likely that the perturbation analysis is still valid.

We now find the changes in action over one motion along the original separatrix to be

$$\Delta \boldsymbol{I} = \varepsilon\Delta\boldsymbol{\delta} \approx (0.002, 0.002) \tag{7.89}$$

Furthermore, in dimensionless units

$$T_{av} \approx 32 \tag{7.90}$$

which needs to be transformed back to physical units. Reversing the time scaling of Section 2 by multiplying by $\sqrt{2/e} \approx 2.2$, we obtain $\delta s = T_{av} \approx 70$. The original time variable t is related to s through the time transformation $\frac{dt}{ds} = r_1 = \frac{e}{2P}(\tilde{u}_1^2 + \tilde{u}_2^2)$. From the unperturbed motion described in Section 3 we find 95 km$\leq r_1 \leq$ 560 km which gives $T_{av} \approx 0.2 \sim 1.2$ yrs.

Thus, a change of $\mathcal{O}(1)$ in I_2 occurs on a time scale of hundreds of years. But since bound motions only exist for $I_2^2 \in (0, \frac{64}{27})$ in the unperturbed case, $\mathcal{O}(1)$ changes are likely to lead to unbound motion. Consequently, in the case of Eros, we estimate the depletion due to Arnold diffusion to occur over time scales on the order of hundreds of orbital revolutions, i.e., very short compared to the age of the solar system!

7.7 Perspective

We have reached the highest peak of our journey. The qualitative implications of the mathematical theory were perhaps evident already down by the foot hills. But finally our ascent has been rewarded by a quantitative prediction, albeit modest and incomplete. From our present viewpoint, a landscape of mathematical insight and physical applications reaches to the horizon. We have yet to explore the peak where we stand and the land beyond.

The presence of a complicated web of intersecting stable and unstable manifolds of persistent tori in a restricted three-body-problem with additional radiation pressure has allowed us to estimate the rate at which grains in circumasteroidal orbits would diffuse from essentially bound to essentially unbound motion. Naturally, these estimates should be compared to rates of reinjection of particles flowing from the sun, as well as to the rate of surface dust ejection due to impacts. Similarly, the relevant time scale for diffusion from tightly bound orbits far from the separatrices to nearly unbound motions needs be accounted for. On that note, the events by which particles enter the asteroid vicinity need to be understood since they might indicate that preferred regions of state space exist in which initial states are likely.

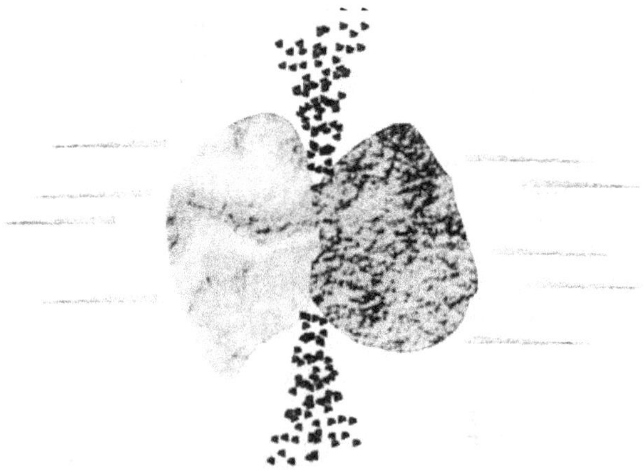

Dust orbiting asteroids might be the results of early collisions. Perhaps only certain regions of phase space are seeded this way.

7.8 Notes and references

The Kustaanheimo-Stiefel transformation introduced in this chapter has been around since the 1960s, but has only recently received well-earned attention. A comprehensive discussion of the transformation and its applications to celestial mechanics is featured in the book of Stiefel & Scheifele[1]. An early application of the transformation to a radiation pressure problem is presented in Kirchgraber[2]. More recently, the present author[3] utilized the KS-transformation to study the stability of certain circular orbits in the radiation pressure problem. Additionally, the three-body problem considered here appears in Dankowicz[4].

We note the work of Hamilton & Burns[5] on the effects of radiation pressure on the long-term stability of grains in orbit about asteroids. In particular, they performed numerical simulations to determine "safe" distances away from the asteroids at which grains would quickly become unbound. The results of this chapter are an attempt towards an analytical foundation for some of the observations of Hamilton & Burns.

1. E. Stiefel and G. Scheifele, *Linear and Regular Celestial Mechanics* (Springer-Verlag, 1971).
2. U. Kirchgraber, "A Problem of Orbital Dynamics, which is Separable in KS-Variables," *Celestial Mechanics* **4** (1971), pp. 340-347.
3. H. Dankowicz, "Some Special Orbits in the Two-Body Problem with Radiation Pressure," *Celestial Mechanics and Dynamical Astronomy* **58** (1994), pp. 353-370.
4. H. Dankowicz, "Escape of Particles Orbiting Asteroids through Separatrix Splitting," to appear in *Celestial Mechanics and Dynamical Astronomy* (1997).
5. D. P. Hamilton and J. A. Burns, "Orbital Stability Zones about Asteroids. II. The Destabilizing Effects of Eccentric Orbits and of Solar Radiation," *ICARUS* **96** (1992), pp. 43-64.

Outlook

At the outset of a journey to a *terra incognita*, one's mind being highly receptive, novelty has its apparent delight. As we become experienced travelers, this initial fascination tends to fade. Instead, new and drastically distinct experiences are sought. At each stop along our path we may stay only briefly, allowing the thrill of the unknown to intrigue and replenish our quest for wonder. On the other hand, those who remain in one place never attain a vision, a larger perspective showing the interconnectedness of all parts of existence. A balance needs to be achieved.

The practical purpose of this final section is to entice the reader and to suggest the inherent incompleteness of any theory including that presented here. It must be emphasized that much work remains. (Although perhaps a complete theory would seem to have but a passing relevance to reality.) Only the persistent student eventually discovers the white areas of the map rather than, as some of his or her fellow travelers, seeking new continents to explore.

It is seldom the case that books of mathematics imply their possible failings. Rather, the first-time reader is often left with the sensation that the material presented was comprehensive and that only slight modifications remain to iron out technical details. With this in mind, the present chapter introduces a few possible avenues for further investigations. Some of these are already being studied by various groups, others have yet to attract attention. Some appear to have immediate implications for the theory within which they arise, others may yield new and unexpected insight into areas not immediately related. This is by no means a complete list. Nor has any attempt been made at ranking these problems in order of importance. My own work in celestial mechanics originated from a similar outlook in a celestial mechanics book. If only one reader of this book might be inspired as I was, then the book has been successful.

Hamiltonian systems

Chapter 3 was devoted to the introduction of sensitive dependence on initial conditions through the existence of exponential splittings and shadowing orbits. The analysis for the two-degree-of-freedom Hamiltonian studied in Chapter 5 was a straightforward extension of these results and thus implied the existence of chaos for this flow. In higher-dimensional problems, however, it is not immediately obvious that the results carry over. In particular, the persistent tori are degenerate structures in that they have several zero eigenvalues associated with them, preventing the standard arguments to be made. Whether chaotic behavior of the type described above is again a consequence of the transversal intersections of the manifolds of the tori remains an open question.

In Chapter 6 we argued that under certain conditions the gaps between persistent tori were sufficiently small to allow the construction of transition chains, and the corresponding shadowing diffusion orbits. A crucial element in the presence of Arnold diffusion of the type described here, the conditions for small gap sizes should be studied in further detail. Also, the calculation of the diffusion rate as described in Chapter 6 should be generalized.

We furthermore argued that Arnold diffusion could take place in the absence of initial hyperbolicity, for example in the neighborhood of tori within the original separatrix. Here the diffusion rates are expected to be exponentially small in ε, and consequently negligible in many applications. While much is known on this topic, there is the expectation that new techniques may bring more detailed insight.

Finally, the study of yet other instability mechanisms in near-integrable systems or other means for state-space diffusion remains very much an open field. Yet further from the present theory lies the analysis of Hamiltonian systems far from the near-integrable case, for which the theory is still in its infancy.

Dynamical astronomy

As a culmination of our efforts, we derived an estimate for escape rates of particles in circumasteroidal orbits. However, only a very limited region in state space was considered. From an astronomical perspective it would be desirable to map out the escape rates for other regions in more detail. Of course, the estimates have limited implications unless processes for replenishing the asteroid's vicinity with grains are considered. We argued that certain creation events, such as collisions, might deposit grains in particular regions of state space, on which subsequent analysis should focus.

One might further imagine other situations such as ring systems where radiation pressure effects prevail. As pointed out in Chapter 7, however, $\varepsilon \sim \sqrt{R}$, where R is the distance from the sun. Consequently, higher-order effects might no longer be negligible at larger distances, thus limiting the validity range of the perturbation.

Also, the effect of diffusion and other state-space instabilities on the formation of the solar system and its constituents should be considered. Dynamical systems theory has only been applied in a restricted number of situations, such as the asteroid belt, where some limited success has been achieved in explaining observed phenomena. No doubt, combining traditional celestial mechanics methods with the mathematical techniques described in this book could provide a highly effective tool.

Finally, on a totally different note, we observe that the equations governing the grain dynamics are very similar to the classical description of the dynamics of a hydrogen atom under the influence of an external electric field. In this context, the present analysis might be extended to apply to calculations of ionization rates of Rydberg atoms in the semi-classical limit. That, however, is a whole other story.

...and he fell silent
and his gaze was lost in the distant...

Index

www.ingramcontent.com/pod-product-compliance
Lightning Source LLC
Chambersburg PA
CBHW050639190326
41458CB00008B/2343
* 9 7 8 9 8 1 0 2 3 2 2 1 4 *